FIRE ATTACK

Strategy and Tactics of Initial Company Response

SECOND EDITION

by

Gary Bush and John Linstrom

STUDENT TEXTBOOK

EMERGENCY WORLD™ LLC

Firefighting is not an exact science, and it involves many hazardous, unknown variables. *Fire Attack: Strategy and Tactics of Initial Company Response* is not intended to provide all the training necessary for safe fireground operations. Emergency WORLD™ will not be responsible for any misunderstanding or misapplication of the information presented in this program. In addition, Emergency WORLD™ will not be liable for any injury or death resulting from the application of information presented in this course.

Chief Content Officer: John Linstrom
Project Manager: Gary Bush
Technical Editor: Susan Peterson
Cover Design: Kear/Stevens Creative

EMERGENCY WORLD™

Loveland, Colorado

ISBN 1-56916-500-9

Printed in the United States of America

Second Edition

Second Printing, December 2002

Contents

1. Understanding the Enemy 1

Introduction 1
Fire Behavior Terms 2
What Is Fire? 3
Types of Fire 7
How Does Fire Move? 7
By-products of Fire 8
Stages of Fire 11
Dangerous Fire Events 13
Construction Types 14
Construction Time Periods 16
Principles Behind Extinguishing a Fire 17
Fire Flow Formulas 18
Summary and Conclusion 19

2. The Basics of Effective Incident Management 21

Command Presence: Leadership on the Fireground 21
Responsibilities of the First-In Company Officer 23
- Radio Size-Up 24
- Command Modes 25
- The Initial Incident Action Plan 27
- Apparatus Placement 29
- Initial Tactical Actions 30

Incident Command System Concepts and Perspectives 31
- ICS for the First-Alarm Incident 33
- Concepts of Tactical Deployment 36
- Situation Status Tracking 39
- Resource Status Tracking 39
- Progress Reports and Tactical Benchmarking 40

Command Worksheet 42
Incident Communications Systems 43
Transfer of Command 45

3. Fireground Risk Management 49

Factors Contributing to Firefighter Injury and Death 49
- Communications Failure 50
- No SOPs/SOGs or Not Following Them 52
- Lack of an Effective Incident Command System 52
- Lack of an Effective Personnel Accountability System 53

Using a Fireground Accountability System 54
Two-In/Two-Out Rule 56
Rapid Intervention Teams 57
Monitoring the Structural Environment 59
Reducing Risks 60
Summary 62

4. Tactical Deployment of Engine and Truck Company Operations 63

Engine Company Operations 63
- Role of the Engine Company 63
- Fire Attack Options 64
- Primary Engine Company Functions 65
- Locating the Fire 65
- Confining the Fire 65
- Extinguishing the Fire 66
- Overhauling the Area 66
- Securing the Scene 66
- Safety During Interior Operations 66
- Tactical Placement of Engine Companies 67
- Size of Attack Line 67
- Nozzle Pressure 68
- Dual-Pressure Automatic Nozzles 69

Truck Company Operations 70
- The Role of Truck Companies 70
- Basic Functions of Truck Company Operations 71
- Response and Placement 72
- Forcible Entry 72
- Occupant Search and Rescue 73
- Utility Control 74
- Placement of Ground Ladders 75
- Ventilation 75
- Extension Check for Hidden Fires 78
- Salvage of Unburned Contents 78
- Placement of Aerial Devices 79
- Provision of Elevated Fire Streams 80

Seat Assignments for Truck Company Responsibilities 80
Summary 82

5. Standards of Fire Attack: Offensive Operations 83

Offensive Fire Attack 83
Offensive Considerations 84
Managing Searches 85
- Search Strategies 85
- Types of Searches 85
- Responsibilities of Command 87
- Mission of Search 87
- Where to Search 88
- Limiting the Primary Search 88
- Maintaining the Focus of the Search 89
- Search vs. Rescue 89
- Hoselines for Search Teams 90
- Backup Line 90
- Searching With a Tool 91
- Searching With a Thermal Imager 91
- Searching for Life or Fire 91
- The Oriented Search 92

Ventilation 93
- Ventilation Theory 93
- Ventilation Tactics 95
- Mechanical Ventilation 96
- Positive Pressure Ventilation 102
- Negative Pressure Ventilation 106
- Confined Space Emergencies 112
- High-Rise Buildings 116
- Below-Grade Conditions 118
- Other Situations 119
- Air Movement for Cooling 119
- Equipment Use Instructions 121
- Safety Precautions 123
- Fan Maintenance 124
- Other Uses for Fire Service Fans 124

Salvage and Overhaul 126
Summary 126
Student Exercises 128

6. Standards of Fire Attack: Defensive Operations 137

Initiating Defensive Fire Attack 137
Concept of Defensive Operations 139
Marginal Fire Attack 140
Transition of Strategy: The Operational Retreat 141
Tools 143
Coordination of Division/Group/Sector Officers 144
Summary 146
Student Exercises 148

Index 155

About the Course Instructors

Gary Bush
Line Battalion Chief
San Bernardino County (CA) Fire Department

Chief Gary Bush serves as a Line Battalion Chief for the San Bernardino Fire Department, which is comprised of 4 field divisions, 9 field battalions, and 63 fire stations. He has been a credentialed community college instructor for more than 10 years, delivering courses in fire service leadership, management, and command. He has also served as a fire academy instructor for more than 10 years, providing training in truck company operations.

John "Skip" Coleman
Deputy Chief
Toledo (OH) Fire Department

Chief John "Skip" Coleman has served with the Toledo Fire Department for more than 27 years. He has published two books on fire management, *Incident Management of the Street-Smart Fire Officer* and *Managing Major Fires*, and he is an Editorial Advisor for *Fire Engineering* magazine. He is also a member of the FDIC Educational Advisory Committee and is a Resident Instructor at the National Fire Academy for courses on incident command and disaster management.

John Linstrom, E.F.O.
Chief Content Officer, Emergency WORLD™

John Linstrom has served more than 20 years in fire departments and EMS agencies in both California and Texas. He has held ranks from firefighter/paramedic through Assistant Chief. He is a National Fire Academy Executive Fire Officer graduate and adjunct instructor. In addition, he is an active member of the Federal Mass Fatalities Recovery Team (FEMA Region IX), as well as a member of the California Task Force 6 of the National Urban Search and Rescue Response System.

John Mittendorf
Battalion Chief
Los Angeles (CA) City Fire Department

Chief John Mittendorf has more than 30 years of experience with the Los Angeles City Fire Department. His background includes extensive experience as an international lecturer on tactics and strategy, and he is the author of numerous fireground operations articles for fire service magazines in the U.S. and Europe.

Michael L. Smith
Deputy Chief/Director of Safety
District of Columbia Fire Department

Chief Michael L. Smith has spent more than 28 years with the District of Columbia Fire Department. He is an adjunct instructor at the National Fire Academy and an instructor of building construction and advanced strategy/tactics at the University of the District of Columbia. His education includes degrees in fire science, construction management, and labor law, and he holds a journeyman carpenter certificate.

Before you begin . . .

This student textbook is designed to be used with the videotape series, *Fire Attack: Strategy and Tactics of Initial Company Response*. Follow these simple procedures as you view the videotapes and use this manual:

- This course is best facilitated in a group situation under the supervision of your agency's trainer.
- An instructor guide is available from Emergency WORLD™. The instructor guide provides lesson plans, exams for each module, overhead transparency masters, and other information. It is a valuable resource for the course facilitator.
- For best results, read the corresponding textbook pages before viewing the videotapes.
- When viewing the videotapes, you can follow the general outline used by the video instructors by observing the main topics in the textbook.

Acknowledgements

In our efforts to research and write a strategy and tactics manual for fireground operations, we discovered a wealth of human resources that made developing this program a less formidable process. The stimulation of working alongside some of the best and brightest minds in fireground operations helped spur us on during the months that were needed to update the course for the Second Edition.

We want to acknowledge the instructors we worked with, who faithfully offered their talents, experience, and professional guidance in spite of their own demanding schedules. Each instructor deserves applause for patience in dealing with numerous phone calls and ongoing requests for ideas, examples, and suggestions regarding the textbook and instructor guide. In alphabetical order, the instructors are:

Alan Brunacini, Phoenix, AZ
Peter Bryan, Rancho Cucamonga, CA
Dave Burns, Valparaiso, IN
Gary Bush, Apple Valley, CA
Tracy Carleton, Spring Valley Lake, CA
Chris DeWolfe, Dover, NH
John "Skip" Coleman, Toledo, OH
Denis Compton, Mesa, AZ
John Linstrom, Apple Valley, CA
Richard Marinucci, Farmington Hills, MI
Robert May, Big Bear Lake, CA
Ernest Mitchell, Pasadena, CA
John Mittendorf, Medford, OR
William Peterson, Plano, TX
Steve Ruda, Los Angeles, CA
Tom Ruzich, Julian, CA
John Salka, New York, NY
Michael L. Smith, District of Columbia
Mario Trevino, San Francisco, CA
Bruce Varner, Carrollton, TX

We are also grateful to all the departments that assisted us in the development of this program, from providing digital images to participating in demonstrations and supporting us in a myriad of other activities. These departments provided us with much-needed background information and examples upon which to base our research. In alphabetical order, the contributing departments are:

Apple Valley Fire District, CA
Big Bear Lake Fire District, CA
Federal Emergency Management Agency
Phoenix Fire Department, AZ
Rancho Cucamonga Fire District, CA
San Bernardino County Fire Department, Station 22, CA
Toronto Fire Services,Toronto, Canada
Victorville Fire Department, CA

Behind the scenes is the editorial staff, who worked long and hard to make this program happen. A special thanks is extended to Mark Sprenger, the founder of Emergency WORLD™, for his unwavering vision, foresight, and leadership. Without his guidance, we couldn't have accomplished our goal of developing a comprehensive, turnkey program for fireground operations. We also owe much gratitude to John Linstrom, the production coordinator and co-author, for his dedicated production direction, and to Chief Gary Bush, co-author and project manager, for his tireless efforts at researching, writing, and managing this project. We also want to thank Susan Peterson for the many hours she spent reviewing and formatting the material.

This fire attack program could not have been completed without all of these participants.

MODULE

1

Understanding the Enemy

Objectives

After completing the reading and activities in this chapter, students should be able to:

OBJECTIVES

- Define fire behavior terminology.
- Define fire.
- Describe the fire tetrahedron.
- Describe the different types of fire.
- Explain the methods of heat transfer.
- Describe the by-products of fire.
- Describe the stages of fire.
- Differentiate flashover and backdraft.
- Describe the five building classifications.
- Describe the methods of extinguishment.

Introduction

When recruits enter the military, they are not just pushed into combat. First, they must go through a series of lessons in basic training. Training is given on military history, chain of command, tools of the trade, strategy, tactics, and many other necessities. After this basic training has been completed, recruits become members of a team and begin preparations with that group before being launched into battle. Each recruit's unit will work together, train together, and go into battle together.

Once recruits have completed the basic training and are working with a unit, additional training will occur, and more tasks/responsibilities will be reviewed and practiced. When a unit is activated for a mission, the commander will require some basic information to get started, including the type of mission, the assigned resources, the goals/objectives, and—perhaps the most important—reconnaissance or information about the enemy.

Firefighters have a great deal in common with the military. Most firefighters go to boot camp (fire academy/recruit school). After graduation, they are assigned to a unit (company) and receive additional on-the-job training from the company officer and other members of the company. When the alarm sounds, the company comes together and functions as a unit.

When responding to an alarm, firefighters receive some reconnaissance about the position, strength, and location of the fire. This information can be received from the communications center that questions the caller about the situation. The company officer and members of the company can reflect on previous incidents, on their experience, and on their knowledge of their response district to acquire important reconnaissance information about the incident.

As a firefighter, you should have a solid understanding of fire and of fire behavior, so that you don't get surprised on the fireground. Let's review some of the basic fundamentals of fire behavior in our quest to win the battle against fire.

Fire Behavior Terms

Understanding the enemy is important when going into battle. It includes understanding how the enemy lives, eats, sleeps, and behaves in certain situations. Fire doesn't live, eat, or sleep, but it does have behavioral characteristics that can be observed and learned. Combustion is a process that follows the laws of chemistry and physics. Understanding the applicable concepts, laws, and terminology will help you understand the enemy and be more prepared for the battle. Following is a review of some basic fire chemistry terms and how they may influence your knowledge of the enemy.

MEASURING TEMPERATURE

- **Heat** – A form of energy that raises temperature.
- **Fahrenheit** – A temperature measurement of heat, where water boils at 212°F and freezes at 32°F.
- **Celsius** – The metric equivalent of the temperature measurement of heat, where water boils at 100°C and freezes at 0°C.

$$212°F = 100°C$$
$$32°F = 0°C$$

Understanding the two temperature scales provides a basis for understanding what may be considered dangerous in regards to extreme temperatures. Also, certain hazardous materials have physical characteristics that use the temperature scales to quantify the physical properties of the materials.

MEASURING HEAT RELEASED WHEN A PRODUCT BURNS

- **British thermal unit (BTU)** – A measurement of heat. 1 BTU is equivalent to the amount of heat needed to raise the temperature of 1 pound of water 1 degree Fahrenheit.
- **Calorie** – The metric equivalent for the measurement of heat. 1 calorie is equivalent to the amount of heat needed to raise the temperature of 1 gram of water 1 degree Celsius.

The BTU and the calorie are used to measure the amount of heat that will be released when a product burns.

FUEL/OXYGEN RANGE THAT ALLOWS COMBUSTION

- **Flammable range (explosive limit)** – The range of proportion of fuel to oxygen that will allow combustion. For example, acetylene will burn in a mixture ranging from 1% to 99%. Natural gas will burn in a mixture ranging from 5% to 15%.The flammable range is important in the understanding of fires involving flammable liquids and gases. If there is too much fuel in proportion to the amount of oxygen, the mixture will not burn because it is too rich. Conversely, if there is not enough fuel to oxygen, the mixture still won't burn because it is too lean. Don't be fooled when the mixture is too rich and won't burn, because at some point, as the conditions change, the mixture will be in the right proportions for burning.

It is recommended that you review your combustible gas meters for measurement scales. Most of the meters will measure the lower flammable range of a material. Knowledge of the meters and measuring scale can help keep you from being exposed to an explosive atmosphere.

- **Flash point** – The minimum temperature at which a liquid material will release sufficient vapors and create an ignitable mixture at the surface. This is a measurement scale used to check the volatility of a material. The lower the flash point, the easier a material will burn.

The lower the flash point, the easier a material will burn.

- **Ignition temperature** – The minimum temperature to which a material must be heated to support combustion. As with the flash point, the ignition temperature can help you understand the potential volatility of a material.
- **Vapor pressure** – A measure of a material's ability to evaporate. The higher the vapor pressure, the greater the chance for the material to travel and find an ignition source.

The higher the vapor pressure, the greater the chance for the material to travel and find an ignition source.

- **Boiling point** – The temperature at which a material's vapor pressure exceeds the atmospheric pressure. At this point, a liquid will begin to change from a liquid state to a gaseous state.

These are just a few basic fire chemistry terms that you should be familiar with. With this baseline of terminology complete, let's jump into more discussion about fire.

What Is Fire?

Simply put, fire is a form of a chemical reaction called oxidation, during which oxygen combines with another element. Some other examples of oxidation are the rusting process and an explosion.

EXAMPLES OF OXIDATION

The rusting of an old truck is an example of a very slow oxidation process. Oxygen from the air combines with the metal on the truck to create iron oxide. This process takes several years to "eat away" the metal into a red-colored mass of easily crumbled metal.

An explosion is on the other end of the scale and is an example of super-fast oxidation. When certain materials are combined, an explosive mixture is created that needs only an electrical charge as a catalyst to begin the reaction. When the mixture is detonated, its components combine with oxygen almost instantaneously and release pressure and heat.

On a scale with rust on one end and an explosion on the other, fire would fall between the two and slightly towards the explosion side (see Figure 1-1). Fire is the rapid oxidation of a material producing heat and light.

Fire is the rapid oxidation of a material producing heat and light.

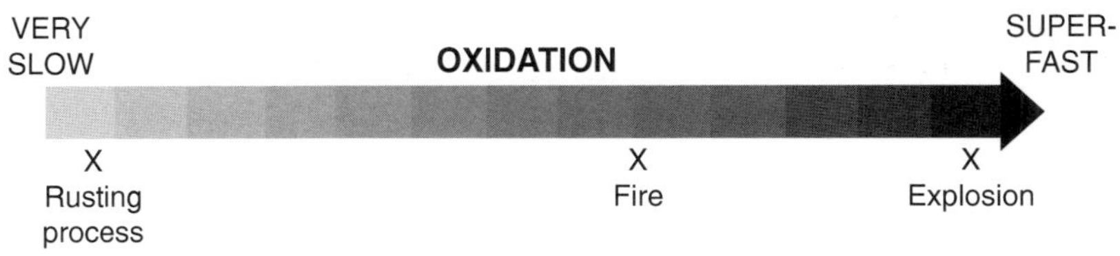

Figure 1-1: The oxidation continuum.

Fire Tetrahedron

The combustion process can be easily explained by discussing the fire tetrahedron. (A tetrahedron is a four-sided, pyramid-shaped object.) The fire tetrahedron consists of four sides, with the sides representing oxygen, heat, fuel, and a chemical chain reaction (see Figure 1-2). The tetrahedron evolved from the fire triangle concept used previously. The chemical chain reaction component added a fourth characteristic to the triangle, making up the three-dimensional tetrahedron.

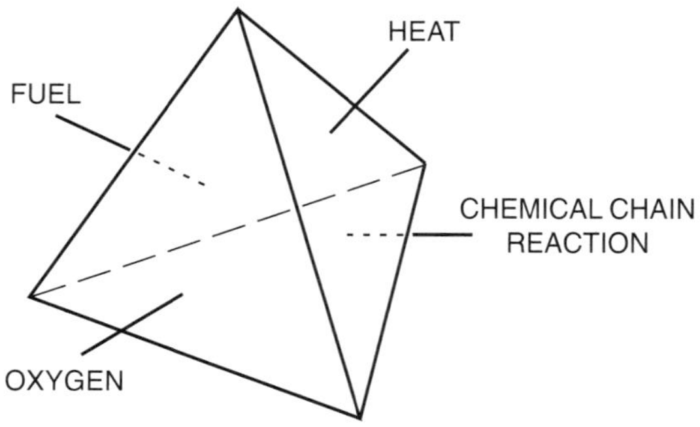

Figure 1-2: The fire tetrahedron.

Oxygen

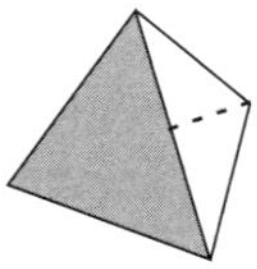

The first side of the fire tetrahedron is oxygen. Oxygen is the most available component in the tetrahedron, since it is all around us in our atmosphere. Certainly, there are areas in our environment that may be oxygen deficient (e.g., confined spaces), but oxygen concentrations are generally around 21% (slightly less in areas of elevation). Air is made up of 78% nitrogen, 21% oxygen, and 1% other gases, including argon, helium, and carbon dioxide.

Fire will burn less efficiently in oxygen-deficient environments and more quickly in oxygen-enriched areas.

Fire will burn when oxygen concentrations are 18-21%. When the concentrations drop below these levels, incomplete combustion will occur with dark, dirty smoke resulting. If a fire is burning in an area that contains oxygen concentrations greater than 21%, the fire will burn with more vigor in this oxygen-enriched area. Simply put, fire will burn less efficiently in oxygen-deficient environments and more quickly in oxygen-enriched areas.

OXIDIZERS

Some materials may burn vigorously because of their chemical structure. These materials are classified, labeled, and placarded as oxidizers. Materials that contain oxygen may burn regardless of the concentration of oxygen in the atmosphere. Learn to recognize the location of these materials. You should anticipate difficulty in extinguishing oxidizer fires, together with higher-than-normal levels of heat. As a rule of thumb, oxidizers end with the suffixes "–ite" and "–ate." Some examples of oxidizers include ammonium nitrate, potassium chlorate, and sodium chlorite.

Heat

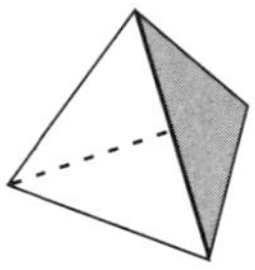

Heat is the second side of the fire tetrahedron. This is the primary point of attack for firefighters. If the heat of a material is reduced below the ignition temperature, then the combustion process can be controlled. Water is the primary resource used in the attempt to reduce the temperature of the burning material.

Heat can be generated from a variety of sources. Understanding each of the sources will allow you to recognize the potential sources of ignition and reduce the opportunity for these sources to come in contact with a fuel source. The sources of heat can be broken into four types: mechanical, electrical, chemical, and nuclear.

SOURCES OF HEAT

Mechanical heat sources

Heat can be generated mechanically by compression and friction.

COMPRESSION AND FRICTION

- **Compression** of a gas causes the molecules to bump into each other more often than normal, thus generating heat. This type of heat can easily be felt in an SCBA cylinder that is charged too quickly.
- **Friction** occurs when one object moves across another. Heat is produced as a result of this motion. Rubbing two sticks together to start a fire is a practical application of friction heat.

Electrical heat sources

Electrical sources of heat may be the most common to firefighters. They include dielectric heating, resistance heating, heat from arcing, static electricity, and leakage current heating.

DIELECTRIC AND RESISTANCE HEATING

- **Dielectric heating** can be seen when a direct or alternating current can be focused on an object. The object is heated when the current strikes the object, stops, and strikes the object again. Dielectric heating is similar to heating that is done in microwave ovens.
- **Resistance heating** occurs when a large amount of electricity is pushed through a small conductor. An extension cord that is supplying too many appliances is an example of resistance heating.

ARCING, STATIC ELECTRICITY, AND LEAKAGE CURRENT

- **Heat from arcing** results when an electrical current "jumps" from one area to another. This heat can be seen in malfunctioning breaker boxes or on a "wires down" response. The heat from the live wires that are providing electrical current can create glass-like materials from sand on the side of roads.
- **Static electricity** results when a positive charge develops on one surface and a negative charge on the other. When the two surfaces come in proximity, the flash and shock will occur. Static electricity can be a dangerous source of ignition and can be found in agricultural and industrial settings. Also, lightning is a type of static electricity and can be the cause in fire ignition.
- **Leakage current heating** can be found on electrical wiring that has had the insulation worn away or damaged. The electrical current will escape out of the wire. The heat generated will break down the affected material, and eventually the material will ignite.

Chemical heat sources

Chemically generated heat happens by methods other than combustion.

MIXING CHEMICALS AND SPONTANEOUS HEATING

- **Mixing chemicals** has the potential for the release of heat. This reaction can be seen when fire streams are applied to water reactive materials.
- **Spontaneous heating** may happen when organic materials are not ventilated properly and heat becomes trapped. The trapped heat will approach the ignition temperature of the object, and once the ignition temperature has been exceeded, the material will begin to burn. This type of heating may occur when oily rags are improperly discarded.

DECOMPOSITION

- **Heat from decomposition** has been the ignition source for barn fires. When wet hay is stored in a barn, the wet grass will begin to break down, and heat will be released. The stacking of hay bales traps this heat. Eventually the loft of hay will ignite, and the material will be consumed in the combustion process. During the early morning hours, heat from decomposition can also be seen in mulch, as the "steam" rises from large piles of bark mulch.

Nuclear heat sources

In nuclear power plants, atoms can be split to create energy. This process is called nuclear fission. The opposite process of combining atoms is called nuclear fusion and is a source of heat.

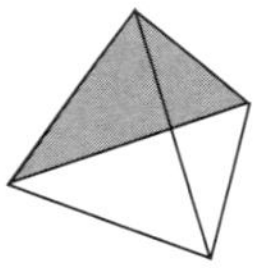

Fuel

The third side of the fire tetrahedron is fuel. Fuels can be divided into three states of matter: solid, liquid, and gas. Before a material can be ignited, it needs to be heated enough to release vapors. When the ignition temperature is reached, the vapors can be ignited. If the material continues to be heated, the fire will grow and will become self-sustaining.

SOLID, LIQUID, AND GASEOUS FUELS

When solid fuels are heated, flammable vapors will be liberated and will ignite. The arrangement of the fuel can contribute to the ignition and spread of fire. If a fuel is finely divided and has a lot of surface area to absorb heat, then the material is easily ignitable. Sawdust will ignite more easily than a log. Also, fire will spread more rapidly in material that is arranged vertically than horizontally. A piece of plywood that is ignited on the ground will burn much more slowly than the same type of plywood that is standing on end.

Like solids, liquid fuels need to be heated to ignite. When heated, the liquid will release vapors that will flash. When these vapors are hot enough and are mixed with the appropriate amount of oxygen, a fire can occur. To reduce the chance of fire, liquids should be properly stored in containers that will not allow vapors to be released.

Gases are ready to burn. They only need a spark to get the fire going! Containers that store gases are designed to prevent leakage, but gas may be released through relief valves if the pressure inside the container exceeds a certain limit. Relief valves are critical components in tank design and are designed to reduce the chances of a tank explosion, or BLEVE, but you should be aware that when relief valves activate and release the product, the fire may grow.

BLEVE

A BLEVE is an acronym for Boiling Liquid Expanding Vapor Explosion. A BLEVE can occur when the internal pressure of a tank exceeds the relief valve's design to relieve pressure, and the inside pressure exceeds the strength of the tank. BLEVEs are an extremely dangerous fire condition. Learn to recognize the potential for BLEVE, and utilize tactics that will reduce internal tank pressures.

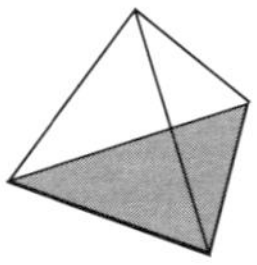

Chemical Chain Reaction

The fourth side of the fire tetrahedron is the chemical chain reaction. This reaction needs to be maintained for the combustion process to continue. The chemical reaction will use the components from the other three sides of the tetrahedron to maintain the reaction and keep the fire burning.

Types of Fire

Fires are divided into four classes, each with a distinct set of characteristics. By identifying the class of a fire, you can select appropriate tools and tactics that will lead to the confinement and extinguishment of the fire.

By identifying the class of a fire, you can select appropriate tools and tactics.

Class A Fires

RESIDENTIAL STRUCTURE FIRES

Class A fires involve ordinary materials. These materials are what you could anticipate encountering within a residential structure fire. Class A materials typically are carbon based and are organic. When the materials are subjected to heat, they release smoke that is dirty gray or black. The flame color is usually a yellowish orange.

Class B Fires

FLAMMABLE LIQUIDS FIRES

Class B fires include flammable liquids. The liquids are usually petroleum based, but alcohol and other types of liquid solvents also fall into this class. The liquids flow to the lowest level and assume the shape of the container. When these liquids are heated, they release vapors that can be ignited. Class B fires usually release large amounts of heat and can be difficult to extinguish. The smoke is usually dark black and is characterized by a bright orange flame. These fires can reignite rapidly and with explosive force if the vapors are not suppressed.

Class C Fires

ELECTRICAL FIRES

Class C fires involve energized electrical equipment. When a fire develops around electrical equipment, take care that you don't use any extinguishing agent that could conduct the electricity and shock you. Once the electricity has been shut off, the fire can then be reclassified to another type of fire.

Class D Fires

FLAMMABLE METALS FIRES

Class D fires occur when flammable metals are ignited. Many flammable metals are water reactive, and applying water to this type of fire can lead to an explosive reaction. The water will mix with the metal, creating corrosive materials and flammable gas. Special encapsulating agents may be needed to extinguish the fire.

Improper recognition of the different classes of fire can lead to improper selection of tactics. If incorrect tactics are implemented, the tasks may lead to unanticipated fire spread and firefighter casualties.

How Does Fire Move?

Controlling heat is one of the fundamental points of fire prevention. To stop a fire, fire protection engineers and code enforcement officials attempt to separate heat and fuel. It is almost impossible to control oxygen content because of the concentration in the atmosphere. So the basis of fire prevention is to separate heat and fuel. Heat can be transferred by a variety of methods. Understanding these methods of heat transfer will allow you to predict the spread of fire and then implement tactics that will impede the fire's progression. Let's take a look at the methods of heat transfer.

The basis of fire prevention is to separate heat and fuel.

METHODS OF HEAT TRANSFER

Convection

Convection is probably the most common method of heat transfer within a structure. Gases that are hotter than their surrounding environment will rise. They will continue to rise until they meet resistance (usually at the ceiling level).

Conduction

Conduction is the transfer of heat through a material. The most common method of conduction is heat being transferred through metal. When metal is heated, it can carry the heat away from the source great distances. Once again, fire prevention efforts will try to separate heat and fuel by trying to eliminate fuel against metal objects.

Radiation

Radiation is the transfer of heat through ultraviolet rays. At the scene of structure fires, the effects of radiation can be seen in many places, as when firefighters exit a building with melted face shields and melted Scott lights. Interior exposures will have "melted" materials about the area. Exterior exposures may exhibit vinyl siding that is damaged, cracked windows, and damaged vegetation.

Applying all these lessons about fire behavior takes time. Seek to develop a keen sense of recognition of what the fire is doing. Don't get caught looking just at what the fire conditions are now. Read the fire, see what it's doing now, and predict where it's going. With this assessment completed, assign resources that will contribute to the control and extinguishment of the fire.

By-products of Fire

LIGHT, HEAT, AND SMOKE

There are three general by-products of fire: light, heat, and smoke. Most fires will produce a flame that will produce light. The light produced usually consists of yellow and orange light. There are some fires, such as alcohol-fueled fires, that will produce a clear flame that is only visible when viewed very closely. The "flames" produced appear to be heat waves.

Heat is a dangerous by-product of combustion. Depending upon the type of material that is burning, various levels of heat can be produced. Different materials that are involved in a fire will release different levels of heat, based upon their chemical structure. Ordinary combustibles of wood and paper release 6,000 to 8,000 BTUs per pound of material. Flammable liquids release 20,000 to 30,000 BTUs for the same weight of material, and 120,000 BTUs are released when a pound of hydrogen gas is burned.

By far, smoke is the most dangerous by-product of combustion.

By far, smoke is the most dangerous by-product of combustion. Smoke not only carries heat, but also contains toxic carcinogens. The make-up of smoke is dependent upon the chemical structure of the material which is burning. Most of the by-products found in smoke are basic chemical compounds mixing carbon, oxygen, hydrogen, chlorine, nitrogen, phosphorus, and sulfur.

More fire deaths occur from carbon monoxide than from any other toxic product of combustion.

Carbon Monoxide

More fire deaths occur from carbon monoxide (CO) than from any other toxic product of combustion. This colorless, tasteless, poisonous, flammable gas is present with every fire, and the poorer the ventilation and the more inefficient the burning, the greater the quantity of carbon monoxide formed. A rule of thumb, although subject to variation, is that darker smoke means higher carbon monoxide levels. Black smoke is high in particulate carbon and carbon monoxide because of incomplete combustion.

The blood's hemoglobin combines with and carries oxygen in a loose chemical combination called oxyhemoglobin. The most significant characteristic of carbon monoxide is that it combines with the blood's hemoglobin so readily that the available oxygen is excluded. The loose combination of oxyhemoglobin becomes a strong combination called carboxyhemoglobin (COhb). In fact, carbon monoxide combines with hemoglobin about 200 times more readily than does oxygen. The carbon monoxide does not act on the body but crowds oxygen from the blood and leads to eventual hypoxia of the brain and tissues, followed by death if the process is not reversed.

EFFECTS OF CARBON MONOXIDE

A concentration of carbon monoxide in the air of about 0.05% can be dangerous. When the level is more than 1%, there is no sensory warning in time to allow escape. At lower levels, headache and dizziness occur before incapacitation, so sufficient warning is possible. The characteristic cherry-red skin color of carbon monoxide poisoning is not always a reliable sign, particularly in long exposures to low concentrations.

Measurements of carbon monoxide concentrations in air are not the best way to predict rapid physiological effects, because the actual reaction is from the concentration of carboxyhemoglobin in the blood causing oxygen starvation. Organs that use a lot of oxygen, such as the heart and brain, are damaged early. The entry of carbon monoxide into the blood will be greater when its concentration in the air is higher. An individual's general physical condition, age, degree of physical activity, and length of exposure all affect the actual carboxyhemoglobin level in the blood.

Experiments have provided some comparisons relating the air and blood concentrations of carbon monoxide. A 1% concentration of carbon monoxide in a room will produce a 50% level of carboxyhemoglobin in the bloodstream in 2.5 to 7 minutes. A 5% concentration in the air can elevate carboxyhemoglobin levels to 50% in just 30 to 90 seconds. Because the newly formed carboxyhemoglobin may be traveling through the body, a person previously exposed to a high level of carbon monoxide may react later in a safer atmosphere. A person so exposed should not be allowed to use breathing apparatus or resume fire control activities until the danger of toxic reaction has passed. Even with protection, a firefighter may become unconscious in a toxic atmosphere.

Administering pure oxygen is the most important element in the first aid care for carbon monoxide exposure.

A hardworking firefighter may be incapacitated by a 1% concentration of carbon monoxide. The stable combination of carbon monoxide with the blood is only slowly eliminated by normal breathing. Administering pure oxygen is the most important element in the first aid care for carbon monoxide exposure. A person may have an uneventful convalescence from a severe exposure, but signs of nerve or brain injury may still appear up to three weeks after the exposure. Again, this is a reason that an overcome firefighter who quickly revives should still not be allowed to reenter a smoky atmosphere.

Hydrogen Chloride

EFFECTS OF HYDROGEN CHLORIDE

Hydrogen chloride (HCl) is colorless but is easily detected by its pungent odor and intense irritation to the eyes and respiratory tract. Although not a general poison, hydrogen chloride causes swelling and obstruction of the upper respiratory tract. Breathing is labored, and suffocation can result. This gas is more commonly present in fires now than it used to be, because of the increased use of plastics such as polyvinyl chloride (PVC) containing chlorine.

In addition to the usual presence of plastics in homes, you can expect to encounter plastics containing chlorine in drug, toy, and general merchandise stores. The overhaul stage of a fire attack is especially dangerous, as toxic fumes linger in a room. Heated concrete can remain hot enough to decompose the plastics in telephone or electrical cable and release more hydrogen chloride.

Hydrogen Cyanide

Hydrogen cyanide (HCN) interferes with respiration at the cellular and tissue levels. The proper exchange of oxygen and carbon dioxide is hampered, so hydrogen cyanide is classified as a chemical asphyxiant. The gas inhibits the enzymes in the tissues which take up oxygen. Hydrogen cyanide also can be absorbed through the skin.

EFFECTS OF HYDROGEN CYANIDE

Materials that give off hydrogen cyanide include wool, polyurethane foam, rubber, and paper. Hazardous HCN atmospheres might be found at fires in clothing stores or rug shops. Exposure to this colorless gas that has a noticeable almond odor might cause gasping respirations, muscle spasms, and increased heart rate, possibly up to 100 beats per minute. Collapse is often sudden. An atmosphere containing 135 parts per million (0.0125%) is fatal within 30 minutes. Nearly all materials tested in an experiment with materials used in the interior of aircraft yielded some hydrogen cyanide.

Carbon Dioxide

EFFECTS OF CARBON DIOXIDE

Carbon dioxide (CO_2) must be considered, because it is an end product of the complete combustion of carbonaceous materials. Carbon dioxide is nonflammable, colorless, and odorless. Free-burning fires generally form more carbon dioxide than do smoldering fires. Normally, carbon dioxide's presence in air and its exchange from the bloodstream into the lungs stimulate the respiratory center of the brain. Air normally contains about 0.03% carbon dioxide. With a 5% concentration, there is a marked increase in respiration, along with headache, dizziness, sweating, and mental excitement. Concentrations of 10%–12% cause death within a few minutes from paralysis of the brain's respiratory center. Unfortunately, increased breathing increases the inhalation of other toxic gases. As the amount of carbon dioxide gas increases, the initially stimulated breathing rate becomes depressed before a total paralysis takes place.

You should anticipate high carbon dioxide levels when a carbon dioxide total flooding system has been activated. These systems are designed to extinguish a fire by excluding the oxygen, and they will have the same effect on humans. According to the American Conference of Governmental and Industrial Hygienists, exposure for even a short time to carbon dioxide concentrations greater than 15,000 ppm should be avoided.

Nitrogen Oxides

EFFECTS OF NITROGEN DIOXIDE

There are two dangerous oxides of nitrogen: nitrogen dioxide and nitric oxide. Nitrogen dioxide is the most significant, because nitric oxide readily converts to nitrogen dioxide in the presence of oxygen and moisture. Nitrogen dioxide is a pulmonary irritant that has a reddish-brown color. When inhaled in sufficient concentrations, it causes pulmonary edema that blocks the body's natural respiration processes and leads to death by suffocation.

In addition, all oxides of nitrogen are soluble in water and react in the presence of oxygen to form nitric and nitrous acids. These acids are neutralized by the alkalies in the body tissues and form nitrites and nitrates. These substances chemically attach to the blood and can lead to collapse and coma. Nitrites and nitrates can also cause arterial dilation, variation in blood pressure, headaches, and dizziness. The effects of nitrites and nitrates are secondary to the irritant effects of nitrogen dioxide, but they can become important under certain circumstances when they cause delayed physical reactions.

Nitrogen dioxide is an insidious gas, because its irritable effects in the nose and throat can be tolerated, even though a lethal dose is being inhaled. Therefore, the hazardous effects from its pulmonary irritation action or chemical reaction may not become apparent for several hours after exposure.

Phosgene

EFFECTS OF PHOSGENE

Phosgene ($COCl_2$) is a colorless, tasteless gas with a disagreeable odor. It may be produced when refrigerants, such as freon, contact flame. It is a strong lung irritant, and its full poisonous effect is not evident for several hours after exposure. The musty-hay odor of phosgene is perceptible at 6 ppm, although lesser amounts cause coughing and eye irritation. A concentration of 25 ppm is deadly. When phosgene contacts water, it decomposes into hydrochloric acid. Because the lungs and bronchial tubes are always moist, phosgene forms hydrochloric acid in the lungs when inhaled.

Stages of Fire

Depending on which school you attended and which book you studied, there have been different names for the different stages of fire over the years. The life cycle of a fire has not changed, but the various stages of fires have been given different names depending on the institution or author. In this course, we will refer to the different stages of fire as the incipient stage, the free-burning stage, and the decay stage.

Incipient Stage

The incipient stage of fire is the very beginning or start of the life cycle of fire. This stage is characterized by a small, visible flame that is confined to the initial material that was ignited. The fire is contained to this initial area of ignition. The oxygen concentrations may have been reduced only slightly. There is a high level of visibility.

Smoke is being generated and is being collected at the ceiling levels. The fire is producing water vapor, carbon dioxide, carbon monoxide, and other toxic gases. Depending on the material, the smoke could be a light gray or a heavy, dark, sooty color.

Even though the flame may be producing a temperature of 1,000°F to 1,200°F, the ceiling temperature is just starting to surpass 100°F. This fire could be extinguished with a handheld, portable fire extinguisher. If the area was protected by an automatic sprinkler system, the fusible link in the sprinkler head would be activated, and the fire would be controlled by this single head. At this point in time, a smoke detector in the immediate area would be activated by the smoke and would provide an early warning to the occupants within the structure.

Key characteristics of an incipient fire are as follows:

CHARACTERISTICS OF AN INCIPIENT FIRE

- First stage in the life cycle of fire.
- Fire is contained to original area and material.
- Oxygen concentrations within the room are relatively unchanged.
- Smoke being produced contains carbon monoxide, carbon dioxide, water vapor, and other toxic gases in small quantities.

- Temperatures at flame are 1,000°F to 1,200°F.
- Ambient room temperature is about 100°F.
- Fire can be extinguished with a portable fire extinguisher.
- Automatic sprinkler system may be activated.
- Smoke detector activates, providing an early warning for occupants.

Free-Burning Stage

If the fire is not controlled by a portable fire extinguisher or an automatic sprinkler system, the fire will continue to grow. It will increase in intensity and will also begin to spread. Smoke and heated gases will rise and move through convection. Adjacent materials will reach their ignition temperature and will begin to burn from the effects of heat transfer by radiation. Depending upon the amount of materials, confinement, and compartmentalization, the fire will enter the free-burning stage. This second stage of fire has the following characteristics:

CHARACTERISTICS OF A FREE-BURNING FIRE

- Second stage in the life cycle of fire.
- Fire is expanding beyond the original area and materials.
- Oxygen concentrations within the room may decrease to about 18%.
- Thick, dark smoke is produced, which contains higher concentrations of toxic gases, especially carbon monoxide.
- Temperatures at ceiling may reach 1,000°F to 1,200°F.
- Ambient room temperature at floor level may be 100°F.
- Fire may be extinguished with a portable extinguisher, but a hose line may be necessary for extinguishment and overhaul.

Decay Stage

If oxygen is added during the decay stage, an explosive reaction will occur with devastating results.

The final stage of fire is the decay stage. Even though this is the final stage, the danger is still high. This stage of fire is evident by plenty of fuel that is heated and smoldering. There are high amounts of toxic and flammable gases at the ceiling level. These gases and the smoldering material are in need of oxygen. The oxygen concentrations within the area have fallen below 18%. If we think back to the fire tetrahedron, this fire has all of the components except oxygen. The chemical reaction is in active progression. There are high temperatures in the area. The fuel is superheated and ready to burn. The only thing missing is the oxygen component. If this oxygen is added, an explosive reaction will occur with devastating results. Vertical ventilation can release the amount of heat at the ceiling level, thus "defusing the bomb that is ready to detonate."

Recognition of the stages of fire and selection of appropriate firefighting tactics will lead to more efficient and effective fire attack and the reduction of firefighter casualties.

Dangerous Fire Events

There are numerous events in the life stages of a fire that can be dangerous. The three most common conditions are flashover, backdraft, and structural collapse.

Flashover

Flashover is a dangerous fire condition in which all of the contents within an area simultaneously ignite. As the fire builds, flashover may occur. Some conditions that may lead to a potential flashover would be:

During flashover, all of the contents within an area simultaneously ignite.

INDICATIONS OF POTENTIAL FLASHOVER

- fire in the free-burning stage
- fire in a contained area or compartment within a structure
- sudden, rapid heat build-up
- tongues or snakes of flames in the overhead
- firefighters being pushed to the floor
- thick, dark, acrid smoke
- highly combustible by-products of combustion that release high concentrations of carbon monoxide
- visible rollover

It is imperative for company officers to understand, recognize, and teach their personnel the warning signs of flashover and the survival skills needed to make it out alive. When a flashover occurs, firefighters will be driven from the area. Those who may be deep in the fire will have only seconds to react and get to a position of safety.

In fighting a fire, always consider the progression of the fire attack. If the conditions exhibit a potential flashover, take immediate steps to position yourself safely. You can use furniture as a barrier, descend stairs, close doors, exit the room, or move around a corner in a hallway to escape the effects of a flashover. Exits should be recognized and cleared if a rapid evacuation is needed.

Flashover can be prevented through ventilation. The methods and tools required for ventilation are presented in Modules 4 and 5.

Backdraft

Backdraft is another dangerous condition that can be fatal for firefighters and can overwhelm a perfectly good tactical plan. Backdraft occurs in a sealed building during the decay stage (third stage) of a fire. In this stage, unburned fuel is still present, all available oxygen has been consumed, and the heat is at a very high level. All that the fire needs to reestablish rapid oxidation at an explosive rate is oxygen. Any failure of a window, door, or unventilated entry due to efforts of an attack team can produce backdraft. In a backdraft, firefighters can be exposed to projectiles and can be thrown by the effects of the explosive force.

Any failure of a window, door, or unventilated entry can produce backdraft.

Fire officers should recognize the following characteristics of an impending backdraft:

CHARACTERISTICS OF AN IMPENDING BACKDRAFT

- pressurized smoke exiting small openings
- thick, dirty, black, gray, or yellow smoke

- confinement
- little or no flame
- excessive heat
- barely audible sounds from within the structure
- pushing and retracting smoke (looks like breathing)
- windows that are smoke stained
- movement of air into any openings that were created by firefighters

The potential of backdraft can be reduced by conducting vertical ventilation to release the heat that has built up at the ceiling level. As noted, ventilation methods will be discussed in Modules 4 and 5.

Structural Collapse

Expect structural collapse after every defensive fire.

The world renowned expert on building construction, Professor Frank Brannigan, has stated that structures are defying gravity from the first day of construction until the day they fall down. Under fire conditions, the progression towards collapse happens even more rapidly. From the onset of ignition until all fire impinging on the structural elements is contained, the potential for structural collapse is present. Company officers must be acutely aware of the fact that the weight of water thrown at the building during extinguishment will adversely affect the stability of the structure. Structural collapse after fire control is a frequent occurrence and should not be a surprise! Expect structural collapse after every defensive fire. A thorough assessment of the structural condition of the building must be completed before assigning crews into the collapse zone to start overhaul or to initiate the cause and origin investigation.

Construction Types

NFPA Standard 220, *Types of Building Construction*, divides the types of building construction into five basic classifications, Type I through Type V. These classifications are further broken down into subtypes by the use of a three-digit code. For example, Type I construction is divided into subtypes 443 and 332. The numbers quantify the fire resistance (FR) of the construction type in hours. The first number is the rating of the exterior bearing walls. The second number is the rating of the structural elements such as columns or girders or other structural elements that support the weight of more than one floor. The third number is the FR rating of the floor assemblies.

The classification types and subtypes are as follows:

BUILDING TYPES

- Type I – Fire Resistive
  - 443
  - 332
- Type II – Non-Combustible
  - 222
  - 111
  - 000
- Type III – Ordinary/Exterior Protected Combustible
  - 211
  - 200

- Type IV – Heavy Timber
  - 2HH (HH refers to heavy timber members. Specific dimensions are further called out in NFPA 220.)
- Type V – Wood Frame
  - 111
  - 000

Type I

Type I construction is referred to as "fire resistive" construction. It is the most stringent type of construction. The structural members are constructed of only non-combustible or limited combustible materials and are also required to be protected by specific construction processes or features. NFPA 220 subdivides Type I construction into two additional types, Type I (443) and Type I (332). In all Type I construction, the protection of the structural elements defines the subtype of construction. An example of Type I construction would be a high-rise building with a steel frame protected by concrete.

FIRE RESISTIVE CONSTRUCTION

Type II

Type II construction is referred to as "non-combustible" construction. It is characterized by non-combustible or limited combustible materials used as the structural members. Like Type I construction, the subdivisions of Type II construction are based on the fire rating (in hours) of the structural elements. The non-structural building materials are also required to be non-combustible. The structural elements are not required to be protected by concrete, spray-on coatings, sheetrock/gypsum board, or other protection, as would a Type I structural element. Both Type I and Type II are non-combustible buildings; however, the Type I building is more resistive to fire because of the protection of the structural members. A common example of Type II construction is a large home improvement chain or a strip commercial retail center with unprotected steel columns, a metal truss roof system, and a metal roof deck. The subdivisions of Type II construction include 222, 111, and 000.

NON-COMBUSTIBLE CONSTRUCTION

Type III

Type III construction was once referred to as "ordinary" construction. NFPA now refers to Type III as "exterior protected combustible." Exterior walls are of limited or non-combustible materials. The standard pre-1933, unreinforced masonry structures in most "downtown" or "old town" sections are of this type. Interior walls, columns, and floors are nearly always constructed of wood. Some ceiling assemblies are commonly made of lath and plaster or metal decking. Type III construction is subdivided into 211 and 200.

EXTERIOR PROTECTED COMBUSTIBLE CONSTRUCTION

Type IV

Type IV is called "heavy timber." This is the traditional "mill construction" that was prevalent in the United States at the turn of the 20th century. Exterior materials must be non-combustible. The interior structural members are non-protected dimensional lumber. These structural members must be specific dimensions to be approved based on the load, size, and span of the building. The interior structural members must be of solid or laminated wood and are not allowed to create any concealed spaces.

HEAVY TIMBER CONSTRUCTION

Type V

WOOD FRAME CONSTRUCTION

Type V is referred to as "wood frame" construction. This construction type is differentiated from Type IV by using smaller-dimensional lumber for construction. The exterior walls are constructed partially or completely of wood. This is currently the predominant residential and light retail construction type in the United States.

Construction Time Periods

It is important to understand not only the types of building construction, but also the major time periods of construction, in order to be better prepared for a fire attack. There are three time periods that should be highlighted: pre-World War II, post-World War II, and the lightweight era.

Pre-World War II

In pre-World War II construction, there was little thought of building in fire-safe features.

The pre-World War II construction type was characterized by custom building using large structural members and true dimensional lumber. Wood was plentiful and often used. Victorian styles with balloon framing (no fire stops) and many concealed spaces were prevalent. Commercial construction included unreinforced masonry (URM), heavy timber buildings, and large, heavy trusses. There was little thought of building in fire-safe features.

In many parts of the country after 1933, Franklin D. Roosevelt's Works Projects Administration (WPA) influenced many commercial construction projects and required reinforced masonry in many government funded projects. Schools, factories, and government buildings erected between 1933 and the end of World War II are usually well supported.

Post-World War II

In the post-World War II time period, a nation returning from war required widespread, affordable housing. The first parkways and freeways were constructed, and large residential tracts made their first appearance. Balloon framing started to become less popular, and heavy timber was less available. Construction was still very solid. The demand for wooden building materials increased. Other types of trusses and new methods to connect buildings together were explored. This time period of well-constructed buildings continued into the early 1970s and 1980s, depending on geography.

Lightweight Era

Buildings built after 1980 should be considered more capable of early collapse until proven otherwise.

The present state of building construction is called the "lightweight era." This lightweight construction era started in different parts of the country at different time periods, but generally between 1975 and 1985. This construction period is characterized by the development of building assemblies: floor assemblies, ceiling assemblies, and framing and truss systems. The weight, girth, and dimension of the structural materials have been replaced by lighter-weight and smaller structural materials and "better" ways to connect these items together. This "better" method improves the cost and the time of construction, but it is more prone to early failure and collapse during fire conditions. Buildings built after 1980 should be considered more capable of early collapse until proven otherwise.

Principles Behind Extinguishing a Fire

There are many tactics that firefighters use to control fires, but there are only four methods that are used to extinguish a fire. Each method corresponds to the removal or reduction of one of the components of the fire tetrahedron (see Figure 1-3). The methods of extinguishing a fire are:

- Reduce the temperature.
- Exclude oxygen.
- Remove fuel.
- Inhibit the chemical chain reaction.

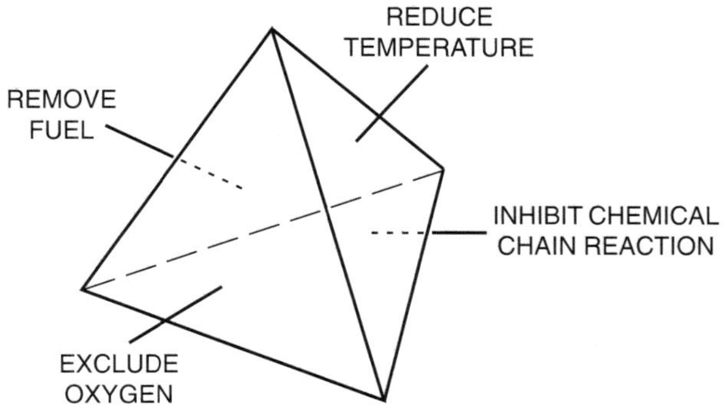

Figure 1-3: The fire tetrahedron, with corresponding methods of extinguishing a fire.

The specific tasks and equipment will be based on the judgment of the officer, the staffing and equipment available, and the current conditions.

Reducing Temperature

The basic method of extinguishing a fire is temperature reduction. Firefighters advance hose into a structure and apply water to the material that is burning, thus reducing the temperature of the material below the ignition temperature. This rapid reduction in temperature knocks out the heat side of the fire tetrahedron and interrupts the chemical chain reaction.

The basic method of extinguishing a fire is temperature reduction.

This reduction in temperature is very effective on Class A fires, but there can be difficulties in using this extinguishing method for other classes of fire. Temperature reduction can be used on flammable liquid fires with limited effectiveness. Water applied to a Class B fire causes the flammable liquid to be sprayed, and the intensity of the fire is increased.

If water is applied to Class C or D fires, firefighters will be subjected to adverse effects. Water applied to energized electrical equipment can lead to electrical shock, and flammable metals can explode when they get wet.

Excluding Oxygen

The exclusion of oxygen works well in Class B and C fires. Class B foam and carbon dioxide are the main extinguishing agents used in these fires. Remember, water should not be used on energized electrical fires, but Class B foam can extinguish large amounts of fire and prevent flashbacks when properly applied.

The exclusion of oxygen works well in Class B and C fires.

When oxygen is separated from the fuel, the oxygen side of the tetrahedron is removed, and the chemical chain reaction is interrupted.

Removing Fuel

Removing the fuel is a viable firefighting option for some fires. Simply shutting off the fuel source by closing a valve is an example of removing the fuel. Also, the physical separation of fuel from fire can be conducted. For example, firefighters could create a firebreak in wildland firefighting or could move pallets of lumber during a lumberyard fire.

Inhibiting the Chemical Chain Reaction

Certain types of extinguishing agents can be applied that will inhibit the chemical chain reaction. Scientists are still studying the chemical chain reaction of combustion, but with the knowledge currently available, multipurpose extinguishing agents have been developed, which are able to interrupt the chemical reaction. When the chemical chain reaction is attacked, the other three sides of the tetrahedron collapse, and the fire will be extinguished.

Try to select the extinguishing method that will extinguish the fire as efficiently as possible with the greatest level of firefighter safety.

It is important to select the most appropriate extinguishing method when attacking a fire. There are pros and cons for each type of fire attack. You should try to select the extinguishing method that will extinguish the fire as efficiently as possible with the greatest level of firefighter safety. This selection takes knowledge of fire behavior and extinguishing principles, together with experience in applying the extinguishing method.

Fire Flow Formulas

There are two recognized fire flow formulas in use by the fire service: the Iowa State University Formula and the National Fire Academy (NFA) Formula.

Iowa State University Formula

CONSIDERATIONS FOR USING THE IOWA FORMULA

This quick fireground formula was the result of testing done in the 1950s at Iowa State University. The typical structural fire loads in the 1950s were mostly organic materials. This was prior to the preponderance of plastics and synthetics in a typical residence or office today. Another factor to be aware of is that ventilation tactics were utilized less often in the 1950s than they are now. Fire attack was more frequently conducted from the outside due to lack of breathing apparatus, protective clothing, and other safety equipment.

The Iowa Formula was one of the first to be widely publicized and practiced. It takes into account the cubic dimensions of an enclosed area to determine minimum gallon-per-minute (gpm) water flow needed to darken the fire in 30-60 seconds. The formula is:

$$\text{gpm} = \frac{L \times W \times H}{100}$$

L = Length
W = Width
H = Height

Let's take a 20 x 20 x 10 room fire. Using the Iowa Formula, one line with minimum of 40 gpm, applied correctly, should darken this room fire down in 30-60 seconds. A second line of higher flow should be deployed as a backup line.

National Fire Academy Formula

The second formula to be developed is the National Fire Academy Formula for Quick Size-up. According to this formula,

$$\text{gpm} = \frac{L \times W}{3}$$

CONSIDERATIONS FOR USING THE NFA FORMULA

The NFA Formula was developed more recently. It takes into consideration the change in residential and light commercial fire load characteristics. A typical fire load today contains more synthetics, which produce more BTUs per pound and increase smoke and noxious gas production. Increased water flow (attack streams) is required due to the increased heat generation. The NFA formula also has adjustments for more frequent ventilation and interior attack tactics that are now typically utilized by contemporary fire departments.

The NFA Formula has been taught in recent years at the National Fire Academy and has simplified calculations to provide more flow for a quicker knockdown. The formula assumes that the floor-to-ceiling height is about 10 feet. The water flow in gallons per minute derived from this formula should darken down the fire in 10-30 seconds. After making the calculation, the company officer can decide the diameter of attack line to select, the number of lines, and the proper flow rate.

As an example, let's take the same 20 x 20 room fire again. One line with minimum of 133 gpm, applied correctly, should darken this room fire down in 10-30 seconds. A second line of higher flow should be deployed as a backup line.

Summary

Company officers must understand the enemy. Fire, heat, smoke, and fire gases can kill. The "battleground" is frequently within an enclosed structure. Understanding the battleground is vital in order to initiate and manage a successful fire attack. Company officers must understand fire behavior and heat transfer in structures. The tactical application of firefighting resources is based solely on this understanding. Officers must also be aware of the building stock in their response area. They must be thoroughly familiar with the type of construction, common construction features, and potential for collapse in order to direct an effective fire attack.

Conclusion

With this review of fire and fire behavior, you should be better equipped mentally to challenge fire. As a firefighter, when the alarm sounds, you are already at a disadvantage because the fire has a head start. However, with your knowledge of fire, of how fire behaves, of building construction, and of the methods of extinguishment, you should be able to make up the difference and win the battle!

Understanding fire behavior is the foundation for building an arsenal for fire attack. Get into the books and review more information about fire. Take these lessons from the textbook and apply them out in the field. Take some time as a company, and look at some of the occupancies within your response district. Put a hypothetical fire in a building and discuss how the fire will grow. Talk about some of the tactics that would be deployed and how the fire would react to these extinguishing measures. When you're on the fireground, remember these lessons about fire behavior. The life you save may be yours!

MODULE

2

The Basics of Effective Incident Management

Objectives

After completing the reading and activities in this chapter, students should be able to:

OBJECTIVES

- Identify the components of and developmental methods for command presence.
- Execute the five basic responsibilities of the first-in company officer to all-risk incidents.
- Utilize the Incident Command System to effectively organize all-risk incidents.
- Utilize command worksheets and incident communications systems as a means for effective command and control.
- Execute a timely transfer of command to an arriving chief officer.

Command Presence: Leadership on the Fireground

Leadership is a word that has found its way into many professional organizations today. For a fire officer, leadership is vital to the daily operation of the fire service, yet it can be quite difficult to exercise, especially during demanding fireground situations. While the majority of this module is dedicated to the effective management of incidents, it is important to highlight the leadership role that all fire officers play while performing incident management. The difference between leadership and management is not always clear.

LEADERSHIP VS. MANAGEMENT

Much emphasis is placed on managing correctly, and rightly so, because good management results in "doing things right." On the other side of the coin is leadership, which is "doing the right thing." The difference between the two approaches is seen in the outcomes. In terms of fireground situations, management (or "command" in street terms) involves selecting the proper strategy and tactical objectives that will result in mitigating the emergency quickly. Leadership, on the other hand, implies the outcomes of influence that the fire officer exudes, with the *people* mitigating the emergency. Simply put, the fire officer's leadership skills have a direct bearing on firefighters who are required to func-

Management is "doing things right." Leadership is "doing the right thing."

Firefighters look to those in leadership positions on the fireground to set the "tone" and "pace" of the incident.

tion in hazardous, stressful, and time-compressed situations. It is during these times that firefighters look to those in leadership positions on the fireground to set the "tone" and "pace" of the incident. This is where we make the connection between leadership and "command presence."

DEFINITION OF COMMAND PRESENCE

Command presence is simply the leadership influence that fire officers should bring with them to every emergency incident—something easier said than done. In a nutshell, fire officers are expected to interject a sense of stability, calmness, and complete control in the midst of chaos. If you are working within a well-designed fire protection delivery system, you will typically be able to provide rapid response time performance to those who need your services. The downside is that such response services put fire officers and firefighters right in the middle of "in-progress" emergencies. This aspect alone brings the stress of rapidly escalating fire problems, rescue problems, and panic-stricken civilians.

To say that those in the fire service are immune to the effects of these dynamics would not be accurate. Yet in the midst of the chaos that can confront us at any hour of the day or night are the fire officers who must step up to the plate and counter the effects of stress. Unfortunately, stress often produces erratic behavior, and fire officers have the ability to wind up firefighters, dispatchers, and the public or to keep them calm and collected in the midst of battle.

Developing Command Presence

The pressing question at hand, particularly for new fire officers, is, "How do I develop command presence?" Here are several methods:

WAYS TO DEVELOP COMMAND PRESENCE

1. *Be mentored by someone who has it*

Experience is a valuable commodity for those who must command emergency incidents. Therefore, it is prudent for anyone desiring to exhibit command presence to seek out those who have more experience than they have. In practical terms, watch how these people operate on the fireground. Notice that they walk and don't run. Watch how they address problems with plans and solutions. Notice their ability to separate themselves from the problems and function in an almost "clinical" manner. Listen to how they sound on the radio. Notice that their radio size-ups for a fully involved dumpster sound the same as for a well-involved multi-story apartment building with imminent rescue problems. And finally, talk to them and glean the lessons that perhaps hundreds of incidents have taught them over the years (i.e., learn through vicarious experience).

2. *Be prepared*

The value of training, drilling, and pre-planning cannot be emphasized enough.

The greatest threat to the fire officer's ability to stay in control comes when something unexpected occurs. The value of training (broadening your knowledge base), drilling (practicing your skills), and preplanning (determining your course of action ahead of time) cannot be emphasized enough. In practical terms, this means addressing all of the occupancies we will discuss in Module 5 and thinking ahead of time about the dynamics of putting fire, smoke, and people into those buildings. Frequent simulation sessions (similar to what airline pilots undergo routinely) become one of the strongest tools for preparing fire officers for the demands that they will inevitably face on the streets, often under the worst of conditions.

3. *Stay physically fit*

Numerous studies in the area of physical fitness have pointed to the value that regular physical exercise brings in reducing stress. Oftentimes in the busy life of an officer, time for physical training is replaced with meeting other demands and responsibilities. At this point, you should realize that stress is your

number one hindrance to developing effective command presence. Steps 1 and 2 above are designed to provide models and strategies for gaining effective leadership influence on the fireground. Nevertheless, one fact remains: we are all human. Both fire officers and those they are responsible for leading are susceptible to the same levels of stress. As a fire officer, it is important for you to be able to "absorb" much of the stress that you encounter. Regular aerobic conditioning is the ingredient for doing just this. If conditioning is done properly and consistently, fire officers will find that the benefits of reduced pulse rate, reduced blood pressure, and proper nutrition will greatly contribute to how they perceive and react to stressful situations—regardless of what hour of the day or night those situations occur.

Stress is your number one hindrance to developing effective command presence.

ABSORBING STRESS

As we progress through the remainder of this module, remember that the common denominator in all incidents is your ability to have a positive leadership impact on the fireground. Your ultimate success in this area will be measured by the degree of command presence you bring to the incident. "Doing the right thing" will result in modeling what your firefighters will admire the most—stability and a calming demeanor.

Responsibilities of the First-In Company Officer

We have just looked at the issues of fireground stress versus command presence. Probably some of the highest levels of stress occur when a fire officer realizes that he or she will be "first in" to a working incident. The reason for this is that the first several minutes of any fire attack operation are typically the most critical. The first-in company officer has the ability either to set the stage for a successful operation or simply to add the arriving company to the problem. It is therefore important to understand five basic responsibilities that fall on the first-in company officer and their relevance to starting a successful attack:

The first-in company officer has the ability either to set the stage for a successful operation or to add to the problem.

1. Conduct a size-up of the incident and transmit that size-up via radio.
2. Select a proper "command mode" as the first-in officer.
3. Develop an initial incident action plan.
4. Place your apparatus and arriving apparatus in the best locations for tactical incident operations.
5. Execute initial tactical actions.

PRIMARY AND SECONDARY GOALS UPON ARRIVAL

The primary goal of each of these steps is to set an incident up so that subsequent arriving companies can easily integrate into the incident and get to work quickly. The secondary goal is to set up an incident that will transition smoothly to the command responsibilities of the first-in chief officer. Frequently, chief officers will have to spend the first 5-10 minutes following their arrival "cleaning up" the tactical errors that have been set in motion, before the remainder of a plan can be implemented. By embracing the following practices, you will always accomplish the primary and secondary goals stated above.

The Radio Size-Up

DEFINITION OF SIZE-UP

The term "size-up" simply means the rapid collection of information that will be needed to formulate an initial incident action plan. The challenge for the first-in company officer is to collect all of the information needed, primarily through visual sources. In Module 1, we learned how to interpret the language that a fire building uses and how to identify all these visual cues. The size-up process for the first-in company officer includes all of these dynamics. The officer must then plug everything into a mental "template" that will quickly result in the intelligence gathering necessary for decision making. In simple and broad terms, three questions must be asked:

QUESTIONS TO ASK DURING SIZE-UP

1. What are the **life safety problems** (i.e., egress of occupants impacted by smoke and/or fire conditions)?
2. What are the **incident control problems** (i.e., what's burning, where's the fire going, and what's in the way)?
3. What are the **property salvage problems** (i.e., what contents in the building stand to be lost due to the fire, and what items stand to be lost due to your operations)?

By addressing these three basic questions, the first-in company officer will be well on the way to formulating an effective initial plan of action.

Now comes the hard part—taking all of the information and boiling it down into four or five sentences, so that you can paint an accurate picture for the remainder of the alarm assignment that is on the way but can't see what you see.

In order to put out an effective size-up on the radio, you should first understand the composition of your listening audience. Assuming that you are the officer of the first-arriving engine company, your "audience" consists of:

AUDIENCE FOR RADIO SIZE-UP

- additional responding engine companies
- responding truck companies
- responding chief officer(s)
- your communications center

With these four categories in mind, remember your primary and secondary goals when transmitting your radio size-up:

- Get arriving companies to work quickly.
- Set the stage for a smooth transition of command for the arriving chief officer.
- A third goal at this point would be to keep your communications center in the loop and postured to anticipate additional resource needs that may develop as the incident progresses.

With these goals taken into consideration, the basic anatomy of a radio size-up looks/sounds like this:

COMPONENTS OF A RADIO SIZE-UP

1. Indication of arrival/confirmation of the address or location:

   "Engine 2 at scene. 1687 Main Street."

2. Description of the fire building:

   "I have a three-story, 50-foot by 200-foot wood frame with two floors of center hallway apartments over a commercial strip."

3. Description of visible conditions:

   "We have heavy smoke showing from one apartment unit on the second floor, middle of the building, and visible rescue from the apartment windows on the second and third floors."

4. Indication of the need for additional resources:

   "Continue the first alarm code 3 and strike a second alarm to report to Level 2 staging at 16th and Main, also anticipating additional alarms for this fire."

5. Indication of initial actions and command mode:

   "Engine 2 will be assuming Main Street IC and setting up for fire attack for the second floor."

With the above five components of the radio size-up in place, let's now look at the positive impacts for the companies responding to your incident:

POSITIVE IMPACTS OF EFFECTIVE SIZE-UP

- Engine companies knowing the location and severity of the fire can now anticipate the "stretch" that will be required to put attack lines into operation. They can also prepare to combat any fire extension into the floor above the fire.
- Truck companies knowing the size and construction of the building can now anticipate the types of ladders that will need to be thrown, the ventilation work that is needed above the fire, and the salvage work that is needed below the fire.
- The responding chief officer(s) knowing the life safety, incident control, and property salvage problems can now be formulating a plan that will augment the first-in officer's initial plan of attack. In addition, issues such as an expanded incident organization and additional resource needs (i.e., tactical, command, and support resources) will be much more quickly considered.
- If your communications center is a busy one, an effective size-up will certainly grab their attention as a working, high-priority incident. Floor supervisors will now be able to posture their personnel to process additional alarms quickly and also to implement fire station move-up plans, in an overall effort to keep resources moving towards your incident and thus reduce response times for greater alarm companies.

Command Modes

Command modes for the first-in officer set the stage for how the incident is initially managed. A direct correlation can be made between the eventual outcome of an incident and the presence of strong, early command in the first few minutes following the arrival of the first-in officer. In order to provide effective incident management, the first-in officer must select and communicate one of three command mode options as part of the radio size-up:

- investigating
- assuming command
- passing command

A direct correlation can be made between the eventual outcome of an incident and the presence of strong, early command in the first few minutes following the arrival of the first-in officer.

Investigating

WHEN TO SELECT THE INVESTIGATION COMMAND MODE

The investigation command mode is used when an insufficient amount of information is available to the first-in company officer from the street or from outside the reported fire building. This command mode is typically selected when there is either nothing showing or only very light smoke showing from the building. In these instances, the officer will typically need to enter the building and investigate the circumstances further in order to complete a proper size-up of the incident. Frequently, the investigation will reveal an unfounded source of an alarm system activation, a small "food on the stove" fire, or steam mistaken for smoke. The investigation command mode allows the first-in company officer to retain all command responsibilities, while providing the mobility to obtain more information. If a working fire is discovered, the first-in company officer may opt either to assume command or to pass command. For a small or previously extinguished fire, the officer may choose to reduce or cancel an incoming alarm assignment.

Assuming command

WHEN ASSUMING COMMAND, IT IS IMPERATIVE FOR THE COMPANY OFFICER TO UNDERSTAND THE IMPORTANCE OF SETTING UP A STATIONARY COMMAND POST.

The second command mode option is to assume command of an incident. This command mode is used when the first-in company officer identifies the need for early coordination and control of incoming companies, coupled with the potential need for ordering additional resources. When this command mode is selected, it is imperative for the company officer to understand the importance of setting up a stationary command post (e.g., in the cab of the officer's apparatus). The command post provides a place where use of communications equipment is optimal. In addition, it provides space for developing a written initial action plan and for tracking the situation status and the resources assigned to the incident.

In order for any officer to complete an appropriate size-up of the fire building, a 360-degree "hot lap" around the structure must be conducted. When an officer assumes command, however, he or she must be disciplined to stay at the designated command post at the conclusion of the size-up process and to maintain command and control of the incident from that location.

It is imperative that the supervision of an officer's assigned company not be jeopardized when the officer assumes command of an incident. Unfortunately, the responsibilities associated with commanding an incident, such as taking a stationary command position, are not compatible with the responsibilities associated with supervising a fire company that is preparing for or making an initial attack on the fire. In these instances, the company officer should exercise one of three "disposition" options for his or her company:

DISPOSITION OPTIONS FOR A COMPANY

1. Field promote the engineer or senior firefighter to assume the role of company commander.
2. Retain the company strictly for purposes of assisting the company officer with command duties (e.g., assist with gathering information about the incident, assist with radio communications, etc.).
3. Transfer the company to the supervision of the next arriving company officer. This is probably the ideal option, in order to maximize the usefulness of the company while placing them under the supervisory expertise of another officer.

Passing command

Passing command is a third option available to the first-in company officer. Exercising this option involves a conscious decision on behalf of the officer, after assessing the size and complexity of the fire problem and its relationship to potential rescue problems. Staffing levels for the first-in company also play a part in this decision. If the officer does not have a sufficient number of firefighters to initiate a fire attack or rescue operation, then consideration must be given as to what initial actions the officer will take (i.e., whether to assist the firefighters or to assume command and control of the incident). The best indication for passing command is a situation in which the first-in company officer's personal participation in an initial attack or rescue effort will most likely stabilize a small fire situation and quickly mitigate a potential rescue problem (e.g., a contents fire in the rear bedroom of an occupied single-family dwelling). Conversely, if the officer's company will most likely not have the ability to stabilize a larger, escalating fire situation, then the officer's efforts are best utilized by taking command of the incident and formulating an initial action plan that will identify the number and types of resources that will be needed to provide rapid mitigation of the emergency.

INDICATIONS FOR PASSING COMMAND

It is important to note that when the first-in officer passes command, it must be clearly communicated as part of the radio size-up. Of equal importance is the fact that the next arriving officer must assume command of the incident and must initiate the remainder of the first-in officer's responsibilities mentioned earlier. It is not an option for the second-in officer to pass command to the third arriving officer and so on. Failure of the second-in officer to assume command will contribute to a fireground operation that is uncoordinated and therefore more hazardous to the firefighters who are required to operate in and on the fire building.

WHEN THE FIRST-IN OFFICER PASSES COMMAND, IT MUST BE CLEARLY COMMUNICATED AS PART OF THE RADIO SIZE-UP, AND THE NEXT ARRIVING OFFICER MUST ASSUME COMMAND OF THE INCIDENT.

The Initial Incident Action Plan

Earlier we discussed the importance of a proper size-up and its relationship to an effective initial plan of action. The plan that is developed by the first-in company officer is the most critical, in that it will set into motion a series of tactical operations that will quickly impact the outcome of the incident, any civilians involved, and the safety of all the firefighters who participate in the plan. In order for a quick and workable plan to be developed, some key considerations must be made:

KEY CONSIDERATIONS FOR THE INCIDENT ACTION PLAN

1. How involved is the fire building?

This key point will set the stage for the first component of the plan. There is a rule of thumb that helps a fire officer quickly calculate the fire involvement noted upon arrival and assists in choosing which of the three fire attack strategies to order. Selecting the proper strategy is critical. For example, selecting a defensive strategy for an otherwise intact building that is 25% involved in fire will most likely result in the unnecessary loss of property. Worse yet, there is the potential loss of people who were still occupying or attempting to exit from the other 75% of the building. Conversely, selecting an offensive strategy for a building that is fully involved places fire companies at immediate risk, due to the inherent likelihood of structural collapse. Selecting the proper strategy also sets into motion a specific set of tactical objectives that reflect the current strategy. Specifics of offensive, marginal, and defensive strategies and subsequent fire attack operations will be discussed in Modules 5 and 6.

IMPORTANCE OF SELECTING THE PROPER STRATEGY

2. Is this a life safety incident?

In specific terms, the fire officer must identify if the fire building is occupied or potentially occupied with "salvageable" victims. If the answer to either possibility is yes, then the following offensive life safety tactical objectives are formulated:

OFFENSIVE LIFE SAFETY TACTICAL OBJECTIVES

- Confine the fire to the area of origin (i.e., keep the fire from spreading to where the people are).
- Search for and remove the occupants (i.e., those who are unable to exit the building on their own).
- Ventilate the structure (i.e., make the environment in the building more tenable for occupants attempting to exit the building and for fire companies attempting to keep the fire from reaching the occupants).

3. Is this a fire control incident?

Once the life safety objectives are identified, then consideration must be given to what will need to be done to address the fire that is involving the contents of the building and/or the building itself. Typical offensive fire control tactical objectives include:

OFFENSIVE FIRE CONTROL TACTICAL OBJECTIVES

- Provide adequate fire flow to effect a knockdown of the fire (i.e., provide a sufficient number of attack lines and gallon-per-minute flow to all areas of the building where active fire is present).
- Provide extension check and extinguishment to all concealed structural spaces (i.e., open up all attic spaces, wall spaces, floor spaces, etc., that are suspected of being involved in the fire and extinguish accordingly).
- Provide subsequent ventilation (i.e., release and/or remove fire gases in direct support of fire attack operations, in both open and concealed spaces within the building).

4. Is this a property salvage incident?

Once the fire control objectives are addressed, the fire officer's attention must then turn to three areas of consideration: the contents that are threatened due to direct fire loss, the contents that are threatened due to indirect fire loss (i.e., as a result of executing life safety and fire control objectives), and the structural load impact of putting a potentially large quantity of water into an already fire weakened building. In order to address these issues, the following three property salvage tactical objectives are needed:

PROPERTY SALVAGE TACTICAL OBJECTIVES

- Protect contents in the immediate fire area of the building.
- Protect contents below any potential accumulation of water from attack lines or sprinkler heads.
- Remove "static" water loads from the building (i.e., water accumulated on nonporous floors and ceilings).

5. How many fire companies will it take to accomplish the life safety, fire control, and property salvage objectives?

Addressing this part of the plan can be somewhat difficult at best, until the full magnitude of the incident can be addressed from both outside and inside the building. However, performing a quick calculation of estimated resource needs is vital in identifying whether the incoming alarm assignment will be capable of executing your plan or if augmented or additional alarms will be required. A sufficient number of companies will be needed to address life safety, fire control, and property salvage tactical objectives in "habitational" type occupancies, such as single-family dwellings and apartment buildings. A sufficient number of companies will also be needed to address fire control and property salvage objectives in buildings where a life safety threat does not exist, such as with commercial building fires that occur after business hours or after all occupants have self-evacuated. Finally, a sufficient number of companies will be needed to address property salvage objectives for fires in all types of occupancies that are out prior to the arrival of the first-in officer (also including the potential for residual smoke removal).

In order to keep fire companies intact for safety and accountability purposes, a good general rule of thumb is to plan for a minimum of one fire company (engine, truck, or rescue company) for each tactical objective identified in your plan. If the incident represents complex rescue, fire control, or salvage problems, then the need for additional companies will have to be factored into your plan for each of those objectives. In addition, it is prudent to address resource needs to staff rapid intervention teams (see Module 3) and to maintain a "tactical reserve" of companies in proportion to the number of companies operating on the fireground, in the event of fireground emergencies (e.g., firefighters trapped) or unexpected developments (e.g., spread of the fire to an exposure building).

Plan for a minimum of one fire company for each tactical objective identified in your plan.

By using this standard approach of identifying the overall strategy and subsequent life safety objectives, fire control objectives, property salvage objectives, and resource needs, the first-in company officer can quickly formulate a prudent and prioritized plan of attack that will greatly assist in carrying out the remaining responsibilities involving apparatus placement and initial tactical actions.

Apparatus Placement

The proper tactical positioning of fire apparatus will set the stage for how effective firefighters will be in carrying out your plan of attack. This is especially true for all first-alarm companies, because initial attack efforts are often heavily dependent on the use of built-in features of the apparatus, such as pumps and preconnected hose beds on engines and aerial devices on trucks. In addition, the proximal location of the apparatus in relation to the fire building provides for rapid access to the wide variety of equipment that fire apparatus bring to an incident.

The proper tactical positioning of fire apparatus will set the stage for how effective firefighters will be in carrying out your plan of attack.

In contrast, placement of greater alarm apparatus tends to be less critical, since their primary mission is to bring additional firefighters and equipment to the scene of an escalating fire emergency. Often, additional apparatus will remain parked in a staging area or spotted outside the immediate operational area. The exception to this typically occurs during defensive fire attack operations, where additional engines may need to be placed to pump and provide required fire flows and where trucks may need to be placed to deliver elevated heavy streams.

The specific guidelines for apparatus placement in relation to certain occupancies will be discussed in greater detail in Modules 5 and 6. The following apparatus placement objectives, however, apply to all incidents. These objectives begin with where the first-in company officer places his/her apparatus, followed by the placement assignments of subsequent arriving first-alarm companies.

APPARATUS PLACEMENT OBJECTIVES

- Place engine companies in the best tactical position that provides for a rapid deployment of attack lines.
- Decide who "owns" the address. In specific terms, look at the vertical height of the structure, and determine whether incoming truck compa-

nies will need to use their aerial devices to accomplish the objectives they have been given. If so, then engine companies must give up the closest spots to the building, in order for truck companies to place their apparatus in the most effective tactical positions. As a rule of thumb, any building greater than two stories in height (with the exception of high-rise buildings greater than eight floors) belongs to the truck companies, when it comes to a limited number of spots that these apparatus can utilize. One fireground truth still remains: you can add more hose to an attack line to make it stretch further, but you can't stretch an aerial device beyond its specified length.

- Place all apparatus in a position that will accommodate the protection of exposure buildings, in the event of a switch in strategy from offensive to defensive operations.
- Place all apparatus outside potential collapse zones, in the event that the fire progresses to full involvement of the structure.
- Place all apparatus in such a manner that they themselves will not become an exposure to radiant heat, etc. Many apparatus have fallen victim to blistered paint, cracked windows, and melted gauges because they were spotted too close to burning buildings.

Initial Tactical Actions

In order to take the proper initial actions, the first-in officer must first put the incident into proper perspective based on the strategy selected.

The final responsibility of the first-in company officer is to put the initial plan of attack into motion. Having quickly formulated the plan, the officer must now decide how and where to use the members of the company in relation to a set of prioritized objectives. It is at this point that many officers can get overwhelmed, since the number of objectives to be accomplished far outweighs the number of firefighters on board the first-in apparatus. Added to this source of stress is the fact that the first several minutes of tactical actions often define the final outcome of the incident. In order to take the proper initial actions, the first-in officer must first put the incident into proper perspective based on the strategy (offensive, marginal, or defensive) selected. The tactical objectives that support this strategy must then be prioritized. It is here that the first-in officer must realize that only one or possibly two tactical objectives will be able to be addressed by the first-in company.

Offensive strategy

INITIAL ACTIONS FOR AN OFFENSIVE STRATEGY

If an offensive strategy will be utilized, the following initial actions will reflect a prudent starting point for executing initial attack operations:

- **For life safety incidents:** Confine the fire (i.e., place the first attack line between the occupants and the fire).

  DETERMINING WHETHER TO CONFINE THE FIRE OR RESCUE VICTIMS

  Note: The first-in officer can be faced with an active fire problem and a simultaneous "visible" rescue problem (i.e., occupants cut off from egress points due to smoke and/or fire conditions), resulting in these victims appearing in a state of extreme panic at windows, balconies, etc. In these situations, the first-in officer must decide whether to commit his/her company to direct rescue operations or to knocking down the fire. A prudent rule of thumb is to attack the "smaller" of the two problems. In practical terms, this means that if the first-in company is faced with a fire situation that they will not be able to confine and/or control, then effecting rescue of the victims is indicated. Conversely, if the first-in company is outnumbered by victims, then an aggressive attack of the fire is in order. If faced with an equally sizeable rescue and fire problem, remember that in most cases the volatility of the incident will rapidly decrease by knocking down the fire or by holding the spread of the fire until more companies arrive.

- **For fire control incidents:** Prepare for a "heavy" fire attack (i.e., secure a water supply and put a large enough attack line or heavy stream in place that is capable of delivering a gallon-per-minute flow sufficient to knock down or hold a large volume of fire).
- **For property salvage incidents (fire out on arrival):** Eliminate sources of flowing water (i.e., open sprinkler heads), and protect contents that are most susceptible to secondary water damage in the immediate area where the fire occurred.

Marginal strategy

If a marginal strategy will be utilized, the following initial actions are indicated:

INITIAL ACTIONS FOR A MARGINAL STRATEGY

- Place attack lines between the involved and the remaining uninvolved portions of the building, and effect occupant rescue.
- Prepare for a defensive attack.

Defensive strategy

If a defensive strategy will be utilized, either of the following initial actions should be used:

INITIAL ACTIONS FOR A DEFENSIVE STRATEGY

- Assess the exposure that is most threatened by the fully involved building and place a large enough attack line or heavy stream between the exposure and the fire building.
- Alternatively, if the involved building can be knocked down by the first-in company, then attack the building. Similar to a life safety incident, the magnitude of the threat to one or more exposure buildings is often most quickly mitigated by knocking down the fire building that is *causing* the exposure problem(s).

Incident Command System Concepts and Perspectives

So far in this module, we have discussed the importance of command presence, and we have identified the basic responsibilities of the first-in company officer. From this discussion, we have realized that effective incident management requires leadership abilities, in addition to the development and implementation of an initial plan of attack. An additional dimension involves the ability to organize the efforts of all the resources that are assigned to an incident, in order to develop a well-coordinated and effective attack on the fire. The best-laid-out incident action plans tend never to see the timely completion of tactical objectives unless fire companies can be integrated into an incident organizational structure that maximizes the efforts of each of these companies. In the absence of organization and coordination on the fireground, a phenomenon known as independent action or "free-lancing" typically occurs. When this is the case, most fire companies will migrate to the most visible and active problem on the fireground (usually where the most fire is) and will vacate other less-than-exciting work to be performed (e.g., salvage).

WHY AN INCIDENT ORGANIZATIONAL STRUCTURE IS NEEDED

A solution to the problem described above is found in the use of the Incident Command System (ICS). This system for the fire service found its origin in the early 1970s, following a rash of wildland/urban interface fires that escalated into major emergency incidents with large commitments of resources from a wide range of fire agencies. Afterwards, a group of the fire service leaders performed an analysis of these incidents and looked at the similarities between fireground operations and the deployment of military combat operations. As a result, many of the same needs for field organization and coordination were identified. An emergency management system was developed, which through subsequent years proved its value in addressing "all-risk" incidents (i.e., structure fires, rescues, haz-mat, etc.), in addition to wildland fires.

An effective Incident Command System includes the following basic concepts:

BASIC ICS CONCEPTS

- A system that connects tactical objectives with tactical resources (i.e., engines, trucks, and rescues to attack the fire, remove the occupants, and ventilate the building).
- A system that connects command and control objectives with command resources (i.e., company officers and/or chief officers to supervise various fireground operations).
- A system that connects logistical needs with support resources (i.e., feeding, fueling, communications, and medical aid for fire companies assigned to the incident).
- A system that allows an incident commander to develop the "size and shape" of the organization in order to implement the incident action plan.
- A system that tracks the location and function of all resources that have been ordered, staged, or assigned to an incident.
- A system that provides for the safety and accountability of all companies operating on an incident through a manageable ratio of one command position supervising a maximum of five companies (ideal ratio).
- A system that provides for integrated fireground communications via face-to-face interaction or via radio communications with all personnel working within the system.

In order for an Incident Command System to be utilized effectively for initial attack incidents, the following perspectives should be considered:

CONSIDERATIONS FOR EFFECTIVE UTILIZATION OF AN ICS

- The size and shape of the incident organization should reflect the size and complexity of the incident.
- The size and shape of the incident organization should expand as the size and complexity of the incident escalate. The "modular" design of the system could, for example, accommodate up to 625 fire companies on a single incident, while adhering to the ideal span-of-control ratio of 5:1 mentioned earlier.
- An incident organization always begins with one officer in charge of the incident who is known as the incident commander or "IC."
- An incident organization is then developed by assigning first-alarm companies into geographical areas on the fireground (e.g., the second floor) or by grouping them together (e.g., a rescue group tasked with remov-

ing all victims from all floors). The assignments are made for the purpose of accomplishing tactical objectives. These functions will be discussed in further detail in Modules 5 and 6.

- An incident organization can then be further developed as needed by assigning greater alarm companies into the existing organizational structure or by adding geographical areas of operation (e.g., floors above the fire floor) or additional groups (e.g., a salvage group tasked with protecting contents on all floors).
- An incident organization can also contract in size and shape as tactical objectives are accomplished and companies are released from the incident.

The ultimate goal for utilizing the Incident Command System is to bring order out of chaos. This is best accomplished by breaking down an otherwise escalating and overwhelming incident into smaller, more manageable parts and delegating the supervisory responsibilities of those parts to capable command officers. When this delegation occurs, the leaders on the fireground have a greater opportunity to exhibit strong command presence, which reduces stress and increases safety.

The ultimate goal for utilizing the Incident Command System is to bring order out of chaos.

ICS for the First-Alarm Incident

As stated earlier, effective management begins with the first-in company officer initiating strong, early command of the incident (unless command is passed to the second-in officer) when confronted with a working fire. A common misunderstanding for users of the Incident Command System, however, results in the majority of tactical resources being used to staff unnecessary command positions within the organization versus being used to mitigate the actual emergency.

A COMMON MISUNDERSTANDING OF THE ICS

A remedy to this misunderstanding is to use the "toolbox" approach to organize initial attack incidents. Each "tool" represents a management or command position in the system. The manager filling each command position will typically be responsible for accomplishing one to two tactical objectives that are contained within your initial incident action plan. An analogy can be made to a vehicle requiring repair. Mechanics have at their disposal a large array of tools designed to assist them in solving mechanical problems. It would be quite cumbersome for a mechanic to dump all of the tools out of the toolbox, since more attention would then be given to dealing with the tools rather than dealing with the problems at hand. Instead, specific tools are taken out of the box, utilized, and then returned to the box. In the same manner, an incident commander will solve each problem by assigning command positions to address each problem and then will terminate those positions as problems are mitigated.

Use the "toolbox" approach to organize initial attack incidents.

The ICS toolbox in the figures that follow addresses only one of four functional sections of the Incident Command System. Command positions within the Operations Section are always the first to be utilized, because this is where tactical operations are conducted. Other sections within the system include planning/intelligence, logistics, and finance/administration functions, which are typically staffed during greater alarm or major emergency incidents, especially those that last for many hours or days.

Using the scenario of a room and contents fire in a single-family dwelling, Figures 2-1 and 2-2 illustrate the typical tools that the incident commander would use and their placement into an organizational structure for a first-alarm incident. Note that the very manageable span of control for both the Incident Commander and the Interior Division Supervisor is 2:1. Also note that due to the simplicity of the incident, all of the other ICS tools remained in the box.

TOOLBOX APPROACH FOR A ROOM AND CONTENTS FIRE IN A SINGLE-FAMILY DWELLING

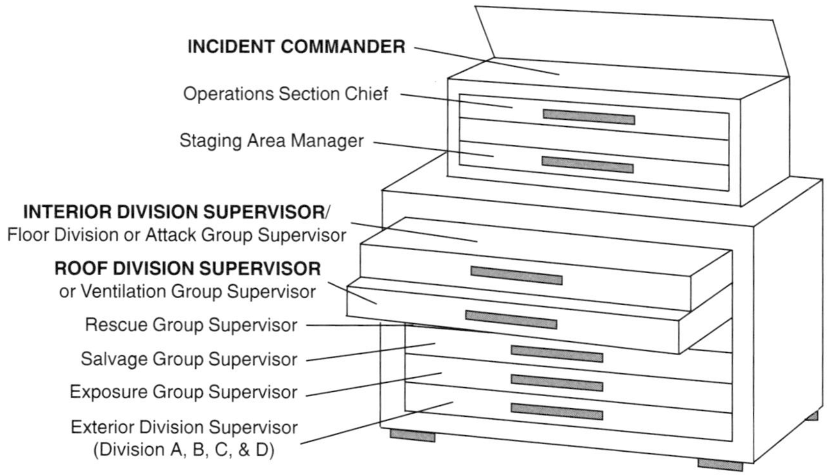

Figure 2-1: ICS toolbox for a first-alarm incident.

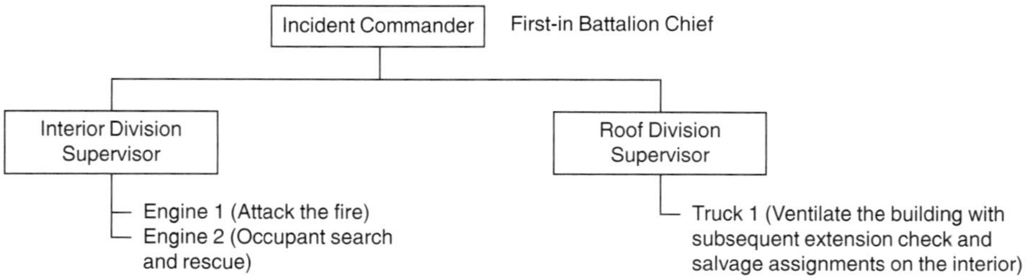

Figure 2-2: ICS incident organization for a room and contents fire in a single-family dwelling.

Using the scenario described in this module in the radio size-up section (see pages 24-25), the next diagram illustrates the tools that the Incident Commander would use for a more complex incident, with expansion capability to accommodate additional greater alarm companies:

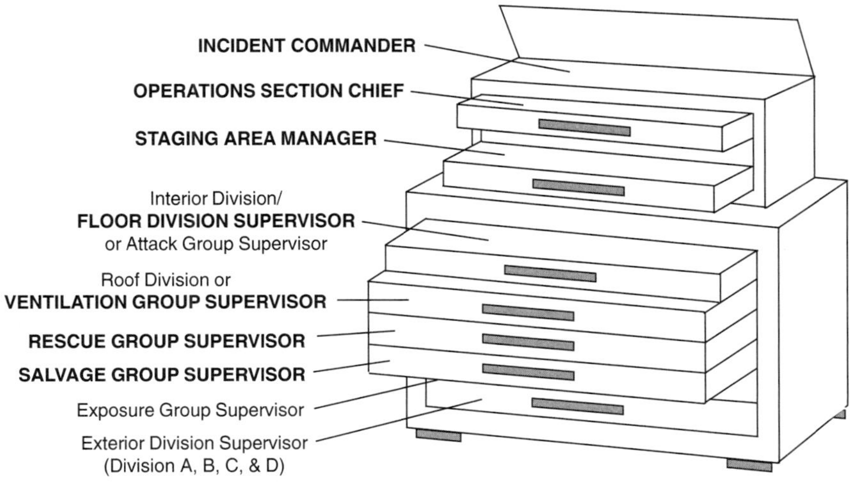

Figure 2-3: Expanded use of the ICS toolbox.

TOOLS FOR A MORE COMPLEX INCIDENT

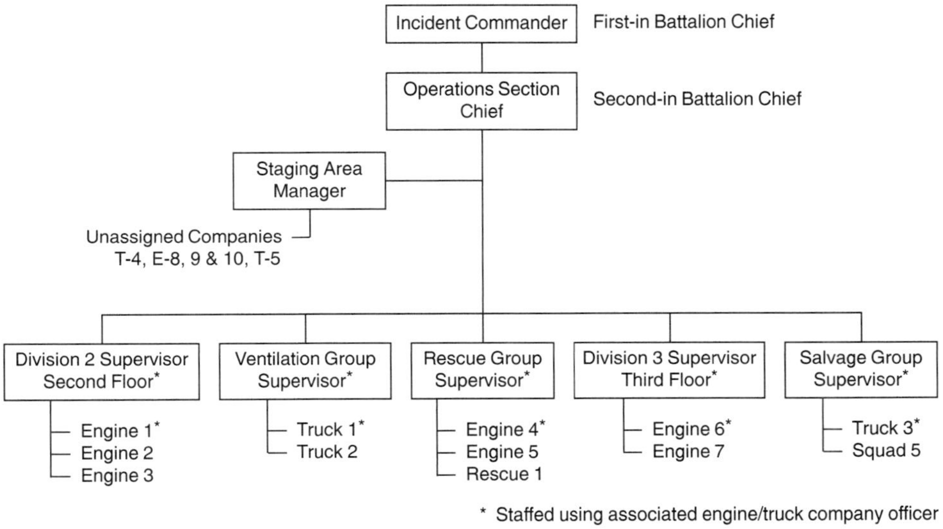

Figure 2-4: ICS incident organization for a more complex incident.

It is important to note that in the above organizational chart, all of the tactical command positions (e.g., Division 2 or Rescue Group Supervisor) are staffed using the company officers that are designated on the chart with an asterisk. Typically during initial attack operations, company officers are tasked with filling these command positions while also supervising their respective companies, unless they disposition their company as was discussed on page 26. Using this approach allows for the rapid deployment of fire companies to mitigate the emergency while simultaneously organizing and coordinating all fireground activities. In addition, in this organizational scheme, a chief officer is assigned as the Operations Section Chief to oversee all divisions and groups operating on the incident, with a span of control of 5:1. The Operations Section Chief is assisted by a Staging Area Manager (typically the company officer who arrives first to the staging area), who manages all unassigned companies.

Concepts of Tactical Deployment

DEFINITION OF TACTICAL DEPLOYMENT

The term "tactical deployment" simply reflects an incident commander's ability to put a typical assignment of first-alarm companies to work on the fireground rapidly. The progression of most structure fires during an initial attack phase is usually predictable, thus allowing an incident commander to formulate a plan, organize the incident, and plug companies into that organization quickly.

Typical deployments

For many departments, the tactical deployment of first-alarm companies is predesignated, while leaving flexibility to address unusual situations. Assuming a typical first-alarm assignment of two engines, one truck, and a chief officer, an example of a typical deployment for a single-family dwelling would be:

TYPICAL DEPLOYMENT FOR A SINGLE-FAMILY DWELLING

- First-in engine: Perform size-up, assume command, and attack the fire.
- First-in truck: Effect inside and outside truck operations (see Module 4).
- Second-in engine: Lay a supply line and provide a backup attack line.
- First-in chief officer: Transfer command from the initial incident commander, develop a tactical action plan, and develop an incident organization to reflect this plan.

Assuming a first-alarm assignment of four engines, two trucks, and two chief officers, a typical deployment for a large commercial building, with no rescue problems, would be:

TYPICAL DEPLOYMENT FOR A LARGE COMMERCIAL BUILDING

- First-in engine: Perform size-up, assume command, lay a supply line, and attack the fire.
- First-in truck: Shut off utilities and ventilate the building.
- Second-in engine: Lay an additional supply line and provide a backup attack line.
- First-in chief officer: Transfer command from the initial IC. Develop a tactical, command, and support plan for the incident, as well as an incident organization to reflect this plan.
- Second-in truck: Throw additional egress ladders, assist the first-in truck company with ventilation, or effect inside operations (see Module 4).
- Third- and fourth-in engines: Lay additional supply lines as needed and provide additional attack lines and/or support to the sprinkler systems.
- Second-in chief officer: Assume the Operations Section Chief position and manage the tactical portion of the incident.

The goal of tactical deployment considerations is simply to get fire officers thinking ahead of time as to how they will connect incoming resources to their plan of attack in a timely and efficient manner.

The goal of tactical deployment considerations is simply to get fire officers thinking ahead of time as to how they will connect incoming resources to their plan of attack in a timely and efficient manner. Deployment considerations also prepare company officers and firefighters to anticipate the actions of the other companies in the alarm assignment and how their respective efforts will assist one another in accomplishing a well-coordinated and aggressive attack on the fire.

Staging

While the goal of tactical deployment involves a rapid and systematic assignment of companies, the practice of these arriving resources to hold back from the incident geographically until receiving an assignment is important. This practice is known as "staging," which involves two levels: Level 1 staging and Level 2 staging. In both cases, the intents of staging tactical resources on arrival at an incident are as follows:

PURPOSES OF STAGING

1. Prevent excessive apparatus congestion at the scene.
2. Place apparatus in an uncommitted location close to the immediate scene to facilitate a more effective assignment by the incident commander.
3. Allow time for the incident commander to evaluate conditions *prior* to the placement and assignment of companies.
4. Produce more effective communications by virtue of reducing radio traffic during the critical initial stages of incident operations.
5. Allow the incident commander to formulate and implement an initial incident action plan without undue confusion and pressure.

LEVEL 1 STAGING

Level 1 staging applies to all incoming first-alarm companies, with the exception of the first-in engine, truck, and chief officer. This action allows for the ideal placement of these first two resources (see the sections on the basic functions of engine and truck companies in Module 4), while also providing an optimal location for the arriving chief officer to establish a command post.

In order for Level 1 staging practices to occur in a coordinated manner, many fire departments utilize standard operating procedures (SOPs) that address how incoming companies should initially approach an incident and eventually be deployed.

LEVEL 2 STAGING

Level 2 staging occurs when fires escalate to second alarm or greater incidents. In these instances, a significant number of apparatus and companies may be called to the incident. In these cases, two significant problems can occur:

1. The incident scene can become even more heavily congested with apparatus.
2. The span of control of the incident commander or the operations chief can be exceeded by a large number of unassigned resources.

The solution to the above problems is found in designating an area to which all greater alarm companies can report upon their initial arrival. Examples of such areas include nearby intersections or parking lots that are capable of accommodating a number of apparatus. Designated locations such as these are known as "Level 2 staging areas" that are controlled through the supervision of a staging area manager. This manager serves as the single point of contact for the incident commander or operations chief when a company and possibly their apparatus are required on the fireground (see Figure 2-4 on page 35). Standard operating procedures that also address Level 2 staging practices are included in the sample SOP shown in Figure 2-5.

Standard Operating Procedure
Level 1 and 2 Staging Procedures

1.0 LEVEL 1 STAGING PROCEDURES

1.1 Level 1 staging applies to normal day-to-day alarms which are usually handled by a first-alarm assignment.

1.2 The first-in engine company will respond directly to the scene and operate to their best advantage.

1.3 The first-in truck company will respond directly to the scene and will place themselves to best utilize their equipment and/or aerial device, generally at the front of the building.

1.4 The first arriving Chief Officer will respond directly to the scene and will place his/her unit in an effective command position, generally in front where two (2) sides of the incident can be viewed and at least 200 feet from the incident, if applicable. Remember not to block access of hydrants.

1.5 All other units will stage approximately one block from the scene until assigned by the Incident Commander or Operations Chief.

1.6 Staged units should choose a location providing a maximum of possible tactical options with regards to:

a. Access

b. Direction of travel

c. Water supply

d. Other factors

1.7 Staged units will, in normal response situations, report company designation, their on-scene status, and their geographical relationship to the incident.

"Engine 2 at scene, Staging West at a hydrant."

1.8 Staged units will stay off the radio until orders are received from the Incident Commander or Operations Chief. However, if after a reasonable amount of time no assignment is given, they will inform the Incident Commander of their status.

1.9 Level 1 staging is automatically in effect for all incidents involving two or more units.

2.0 LEVEL 2 STAGING PROCEDURES

2.1 Level 2 staging supersedes Level 1 staging when ordered by the Incident Commander.

2.2 Level 2 staging location shall be designated with the ordering of second or greater alarms. By procedure, the first-in tactical resource to the designated Level 2 staging area (i.e., parking lot or intersection in proximity to the incident) automatically assumes the role of Staging Area Manager. All other incoming resources will report to the designated Level 2 staging area and await direction from the Staging Area Manager.

2.3 Utilization of a Staging Area Manager is a critical span-of-control tool for the Incident Commander or Operations Chief in the management of unassigned resources, particularly during escalating incidents.

Figure 2-5: Sample standard operating procedures for staging.

Situation Status Tracking

Once all initial companies have been deployed, it is imperative for the incident commander to begin to place specific "controls" on the incident. These controls provide the IC with the ability to stay apprised of key events that are occurring on the fireground, in order to minimize risks and therefore maximize firefighters' safety. Every minute that a fire continues to threaten the structural integrity of a building means that an offensive strategy is moving closer and closer to a potential change to a marginal or defensive strategy and subsequent change in tactical objectives. Control methods that the incident commander uses include situation status tracking, resource status tracking, progress reports, and tactical benchmarking.

CONTROL TOOLS

Situation status, or "SITSTAT" in ICS terms, involves tracking the progression of the fire incident in terms of life safety, incident control, and property salvage problems. Tracking these categories of events throughout the duration of the incident allows the incident commander to verify the validity of the incident action plan. A useful tool for tracking situation status is a rough sketch of the fire building that notes the types and locations of problems. Predrawn diagrams such as the one illustrated below are typically one component of a command worksheet that provides space to record this vital information. In addition, pre-fire plans typically contain a predrawn layout or floor plan of the building. (Note: The sample information shown in Figure 2-6 reflects the scenario described on pages 24-25 and the incident organization on page 35.)

ELEMENTS OF SITSTAT

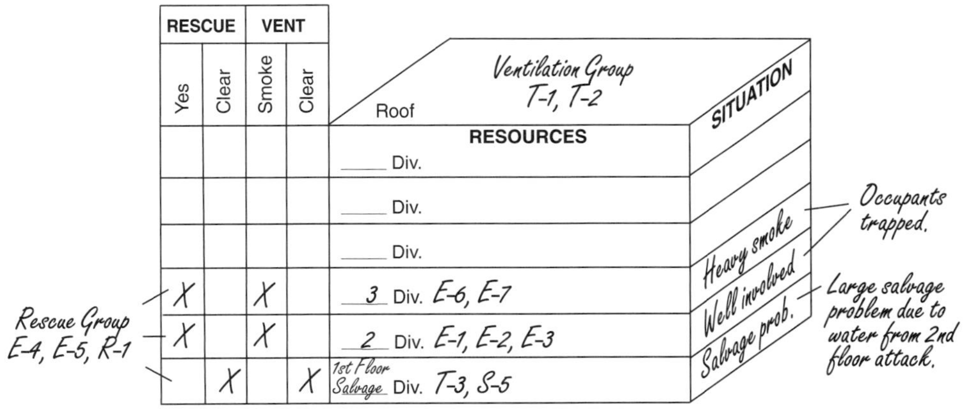

Figure 2-6: Sample situation status tracking form.

Resource Status Tracking

Resource status, or "RESTAT" in ICS terms, serves as the next vital control for the incident commander. Tracking the location and assignment of all companies that are operating on the fireground is critical for the safety of those companies. Fireground emergencies, such as an unanticipated flashover or roof collapse, often require a change in strategy or a rapid accounting of companies. Failure to track companies from the minute they are given an assignment puts their safety in jeopardy, particularly if a change in strategy is executed. For example, when an incident switches from an offensive to a defensive strategy, all companies are withdrawn from inside and from the roof of the building. At this point, the incident commander must account for all companies prior to ordering heavy streams into operation. A thorough RESTAT process makes this a simple task. An incomplete tracking of resources can easily spell disaster for a company still operating inside the building that did not hear the order to withdraw.

FAILURE TO TRACK COMPANIES FROM THE MINUTE THEY ARE GIVEN AN ASSIGNMENT PUTS THEIR SAFETY IN JEOPARDY, PARTICULARLY IF A CHANGE IN STRATEGY IS EXECUTED.

The incident commander must know at all times where his or her companies are located and what they have been assigned to do.

The bottom line is that the incident commander must know at all times where his or her companies are located and what they have been assigned to do. RESTAT also allows the incident commander to view quickly where the largest commitment of tactical resources exists and whether the ratio of companies to division or group supervisors is beginning to exceed a manageable span of control. A simple table as illustrated in Figure 2-7 provides a quick rundown of companies (typically in the order in which they arrived) and is another component of a command worksheet. (Note: The A/S column signifies whether the company has been assigned (A) or is unassigned in staging (S). The locations of the assigned companies are captured in the previous SITSTAT diagram and reflect the incident organization shown on page 35.)

COMPANY	A/S	ASSIGNMENT
1st alarm:		
E-1	A	Division 2 Supervisor
E-2	A	Assigned to Division 2
E-3	A	Assigned to Division 2
T-1	A	Ventilation Group Supervisor
T-2	A	Assigned to Ventilation Group
BC-1	A	Incident Commander
BC-2	A	Operations Section Chief
2nd alarm:		
E-4	A	Rescue Group Supervisor
E-5	A	Assigned to Rescue Group
E-6	A	Division 3 Supervisor
E-7	A	Assigned to Division 3
T-3	A	Salvage Group Supervisor
T-4	S	
S-5	A	Assigned to Salvage Group
R-1	A	Assigned to Rescue Group
3rd alarm:		
E-8	S	
E-9	S	
E-10	S	
T-5	S	

Figure 2-7: Sample resource status tracking form.

Progress Reports and Tactical Benchmarking

As an incident progresses, it is necessary for the incident commander to solicit reports from the key command positions that are tasked with carrying out the incident action plan. In specific terms, the incident commander requires officers in those positions to accomplish the tactical objectives assigned to them. Regular reports as to the progress in accomplishing these objectives are vital in verifying the validity of the plan and in maintaining accurate situation status.

A simple format for officers to utilize in providing progress reports and other related information is the acronym SPIN, which stands for:

FORMAT FOR PROGRESS REPORTS

Situation – What is currently occurring at the officer's location.

Progress – Progress in accomplishing objectives.

Information – Any other intelligence helpful to the incident commander.

Needs – Additional companies needed to accomplish objectives.

HOW TACTICAL BENCHMARKING WORKS

Progress reports lend timely information for the final control tool, known as tactical benchmarking. This process involves looking at all of the key control events in order of typical priority from incident start to finish. This list also serves as a component of a command worksheet and allows the incident commander to track all tasks that have been assigned and completed. In the list shown in Figure 2-8, you will notice that the highlighted items serve as key benchmarks that lead to fire incidents being declared "under control." In other words, the incident commander seeks to accomplish specific things, such as knockdown of the fire, removal of all affected occupants, and extinguishment of any fire that has extended into concealed spaces in the structure. When these benchmarks have been reached, the IC can then declare the incident under control and can concentrate on residual objectives, miscellaneous notifications, and a release plan for the resources at the scene.

- ☐ **FIRE ATTACK**
- ☐ Water supply
- ☐ **PRIMARY SEARCH**
- ☐ **ALL CLEAR**
- ☐ Utilities
- ☐ Ventilation
- ☐ Back-up lines
- ☐ **EXTENSION CHECK/EXPOSURES**
- ☐ Salvage
- ☐ **SECONDARY SEARCH**
- ☐ **ALL CLEAR**
- ☐ Progress reports
- ☐ Rescue laddering
- ☐ Sprinkler support
- ☐ Breathing/lighting support
- ☐ Rehab
- ☐ Traffic control
- ☐ PIO/media notification
- ☐ **UNDER CONTROL**
- ☐ Investigator
- ☐ Utility companies
- ☐ Building representative
- ☐ Building and safety
- ☐ Release plan

Figure 2-8: Sample tactical benchmarking form.

Note: One tracking method for this list is to place half an "X" in the box to indicate that a task has been assigned and to complete the "X" when it has been accomplished. This gives the incident commander a quick view of how the incident is progressing, as well as reminders (especially at 2:00 a.m.) of all the details that need to be addressed while at the incident.

Command Worksheet

One of the most vital tools for the incident commander, other than radio communication devices, is a command worksheet (Figure 2-9). This worksheet enables a fire officer to capture key points of information as an incident progresses, so that accurate and prudent decisions can be made. While many formats of worksheets exist, the anatomy of the worksheet is what's important. Important components that we have discussed in previous portions of this module include:

COMPONENTS OF A COMMAND WORKSHEET

- space to document situation status (SITSTAT)
- space to document resource status (RESTAT)
- space to document tactical benchmarking

Another important component of the worksheet is space to document and track an initial incident action plan. The format shown below triggers important aspects of that plan, to include considerations such as:

1. Strategy.
2. Life safety, incident control objectives, and property salvage objectives. (Note: The format shown allows incident commanders to list specific tactical objectives next to the small check boxes on the left and to list the number of fire companies needed for each objective within the boxes on the right.)

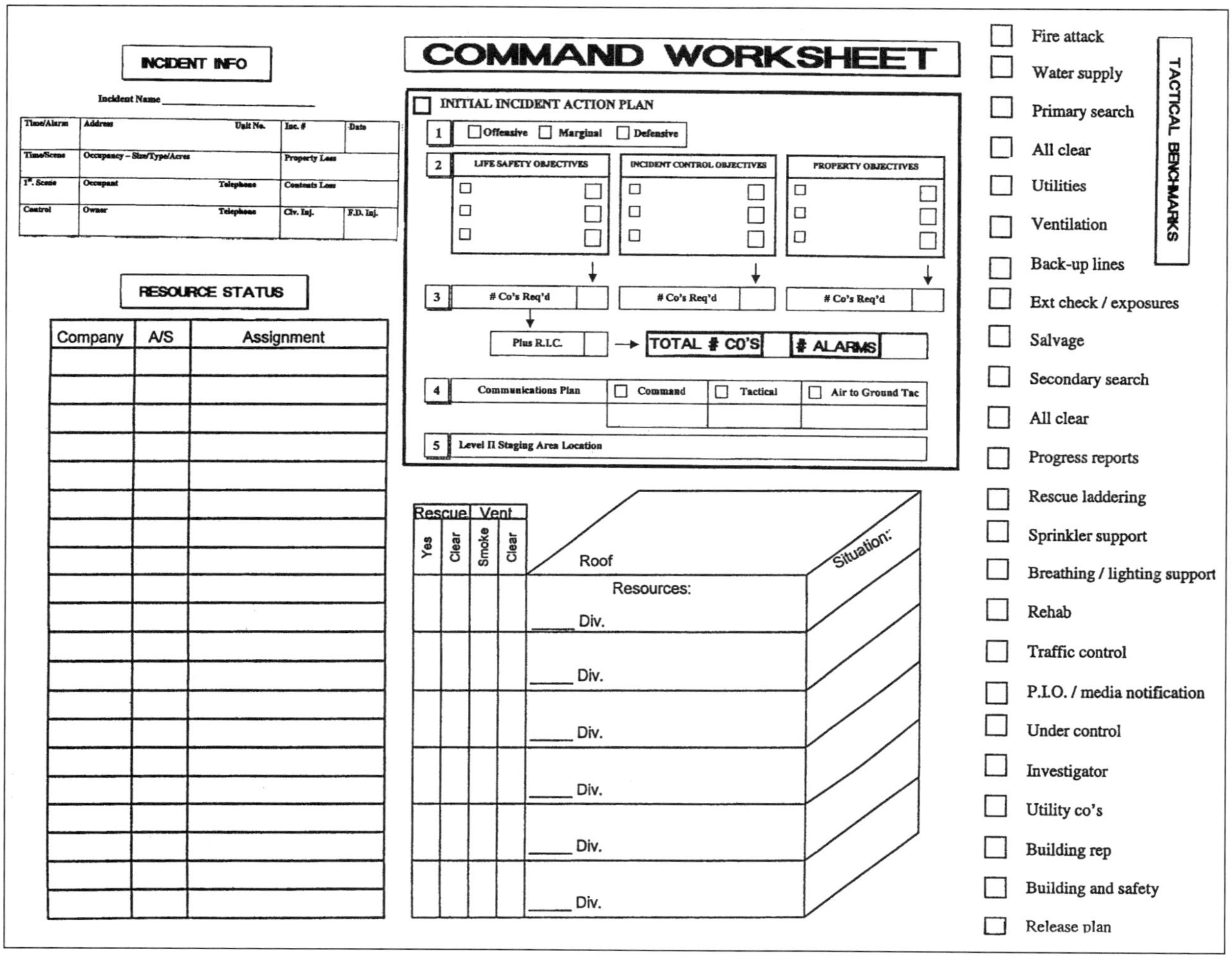

Figure 2-9: Sample command worksheet.

3. Calculation boxes to determine the total number of fire companies required for each category of objectives, to include rapid intervention companies (discussed in Module 3). Based on the number of fire companies contained within each alarm assignment, the IC can then record the number of alarms that will be required to accomplish the plan.
4. Communications planning to include the identification of command and tactical channels (discussed below).
5. Designation of the Level 2 staging area location, to which greater alarm companies will initially report upon arrival at the incident.

You will note that the focal point of the worksheet is the action plan. This plan demands that the incident commander continually assess the plan's validity, in terms of the completion of tactical objectives and the current strategy that is in place.

A final component of the worksheet involves a block to capture specific points of information about the incident, including the time of key events, property/contents loss estimates, and occupant/owner contacts.

BENEFITS OF HAVING A WELL-DOCUMENTED COMMAND WORKSHEET

At the conclusion of this module, we will discuss the aspects of transferring command from the initial incident commander (i.e., first-in or second-in company officer) to the arriving chief officer. Probably one of the most important ingredients to a smooth transition is the ability of the initial incident commander to relay information in a rapid manner and in simple terms. The more that is documented, the less that needs to be spoken. Handing a well-documented worksheet over to the chief officer provides this new incident commander with organized, quick-read items that will enable the chief to build quickly upon the existing plan and prepare for any escalation of incident dynamics.

Incident Communications Systems

Among the most common sources of frustration for the fire officer are problems related to fireground communications. Although you may be miles away from an incident, it is possible to get a pretty accurate sense of how well the incident is going simply by listening to a working fire on the radio. The best-laid-out incident action plan quickly becomes dependent on how well those in command positions are able to communicate among themselves and with the tactical resources assigned to them.

Before we discuss the specifics of incident communications systems, it is helpful to understand some common factors that make communications difficult on the fireground. Such factors include:

FACTORS THAT MAKE COMMUNICATIONS DIFFICULT

- The need to send and receive messages in a rapid manner—this necessitates the use of radios to accomplish a large portion of fireground communications.
- The need for clear understanding of messages that are being sent or received—this ideally is accomplished in a face-to-face manner, where facial expressions and general body language quickly tell the sender if the message makes sense and if the receiver will carry out the orders that are given.
- The need to have adequate radio "air time" to send messages and receive a response to messages—this requires an adequate number of radio channels to be assigned to an incident, in order to minimize the competition for critical air time.

- The need to communicate in the midst of noisy, "combat" conditions—many officers have to send or receive radio messages while crouched next to interior fire attack streams or while on rooftops in close proximity to power saws, and often while wearing an SCBA facepiece.

Considering the above factors that inhibit communications, it is no wonder that incident communications issues are often a "hot" topic during post-incident critiques. In order to manage effective fireground communications, all officers in command positions must constantly assess the effectiveness of the communication process and minimize the above factors.

The following communications practices serve to ensure that all messages, orders, and pertinent information reach all officers in a timely manner:

HELPFUL COMMUNICATIONS PRACTICES

All officers need to utilize face-to-face methods of communication as much as possible.

- All officers need to adhere to strict discipline and protocols when using a radio to communicate. Keep all transmissions short and to the point. Avoid using names and excess verbiage when relaying orders, progress reports, etc.
- All officers need to utilize face-to-face methods of communication as much as possible. This practice applies especially to situations in which lengthy or complex messages need to be given. Examples include task-related orders that occur between division/group supervisors and company officers. Another advantage of face-to-face communications is that they free up valuable air time for those messages that can only be sent via radio.
- All officers need to place themselves in the optimal position possible to communicate. Whether at an outside command post or deep inside a fire building, radio communication can literally serve as the critical link to safe fireground operations. As such, officers must continuously be prepared to receive radio transmissions. Those in command positions on the fireground may find it necessary to distance themselves from the immediate area of operation in order to hear or be adequately heard. In addition, simple tools such as auxiliary microphones for portable radios placed on turnout coat collars provide a more effective means of hearing radio transmissions, instead of relying on a radio speaker muffled inside a deep pocket on the coat. Other tools, such as microphone systems that are integrated into the SCBA facepiece, also make a notable difference in the clarity of radio transmissions.
- The final practice involves a systematic designation and use of radio channels that are managed by both the communications center and the incident commander. Examples of such channel designations include the following:

EXAMPLES OF RADIO CHANNEL DESIGNATIONS

1. Primary dispatch channel – Utilized for the communications center to alert and transmit alarm assignments to fire companies in quarters or out in the field.
2. Incident response channel – Utilized for companies to transmit scene arrival and radio size-ups and for the communications center to provide pertinent information to companies while responding to the incident. Many fire agencies will also utilize this channel designation for incident commanders to use as a "command" channel. As such, incident commanders will use the channel to transmit the need for additional alarms and to give incident updates to the communications center. Likewise, the communications center will use this channel to transmit pertinent information to the incident commander.

3. Fireground tactical channel – Utilized for all officers in command positions to transmit task-related orders to tactical resources. ("Truck 5 from Roof Division, throw egress ladders on the west side of the building.") As can be expected, this is the channel that is the most active in terms of radio transmissions. In addition, an incident commander or operations section chief will also use this channel to transmit tactical-related orders to division/group supervisors. ("Roof Division from IC, ventilate the attic and top floor of the building.") Supervisors will use the channel to transmit progress reports back to the IC or operations chief. ("IC from Roof Division, we have heavy fire extension in the common attic.") If needed, additional fireground tactical channels can be assigned by the communications center as the size and/or complexity of the incident increases.
4. Air-to-ground tactical channel – Utilized for all officers to communicate with any aircraft assigned to an incident. Due to the increased hazard of mixing airborne resources with fire companies operating on the ground, a designated channel must be reserved to ensure rapid and timely radio transmissions between the two. This is especially critical when rotary-wing aircraft (helicopters) are required to land at an incident. Close coordination between the aircraft and the command position responsible for the landing space is heavily dependent on uninterrupted radio transmissions, particularly if a landing needs to be aborted. ("Copter 10 from Roof Division, abort landing! Your tailrotor is about to hit the electrical service lines to the building.")

Use of the above channels in a systematic manner on working incidents serves to facilitate the various needs to communicate. As a result, a high volume of information can be relayed (coupled with face-to-face communication where practical), without the typical competition and confusion that can occur when all players must share the same channel. When used properly, a prudent incident communications plan such as the one described above can simultaneously process a dynamic, escalating incident to include the ordering of additional alarms (response/command channel), the dispatching of additional alarms (primary dispatch channel), the ordering of tactical objectives and related task-level assignments (fireground tactical channels), and receiving air reconnaissance information (air-to-ground tactical channel), without hindering anyone's need to communicate in a timely manner.

Transfer of Command

PASSING COMMAND VS. TRANSFER OF COMMAND

Earlier in this module, we discussed the various command mode options available to the first-in company officer. One option allowed for the officer to "pass" command to the next arriving officer, in order for the first-in officer to participate personally in an initial fire attack or rescue effort. As stated earlier, when this command mode is exercised, then the next arriving officer must assume command of the incident and will typically retain this position until relieved by the first-in chief officer. When this relief process occurs, it is known as a "transfer of command" and should not be confused with passing command.

Transfer of command is typically indicated when an incident has not been declared under control by the time the chief officer arrives.

Transfer of command is typically indicated when an incident has not been declared under control by the time the chief officer arrives. At this point, the command and control expertise of the chief officer is utilized, in order to continue and/or modify the current action plan and to build upon that plan, in the event that the incident continues to escalate. It should be noted that a transfer of command does not necessarily mean that the company officer in charge was commanding the incident poorly; rather, the company officer may be better utilized to fill a command position on the fireground, provide safety supervision, perform situation and resource status functions, or simply serve as an extra set of eyes and ears for the new incident commander.

When a transfer of command occurs, it is imperative that both the company officer in command and the arriving chief officer are prepared to provide and receive information, respectively, in a timely manner. Unfortunately, the dynamics of a working fire do not cease while the command of the incident is being transferred, nor do the needs and coordination of the companies operating on the fireground cease. Therefore, the initial incident commander must be able to communicate information in an organized manner and in such a way as to allow the chief officer to gain a quick grasp of the magnitude of the incident and the crucial points of the plan that is in motion.

EFFECTING A SMOOTH TRANSFER OF COMMAND

In order to effect a smooth transfer of command, the company officer and the chief officer must first find a place suitable to communicate. The middle of the fireground is *not* a suitable place for this operation. In addition, the arriving chief officer should not have to hunt for the incident commander. Instead, the initial IC should be located in a stationary command position (typically at the front of the fire building), where the arriving chief officer can be quickly postured to meet and converse.

When the actual transfer of command takes place, an organized format of sharing information should be followed. Key points of this format include the following and reflect a typical conversation between the captain on Engine 2 and the first-in battalion chief, from our scenario on pages 24-25 and the initial phase of the incident organization on page 35:

FORMAT OF SHARING INFORMATION

- Current situation – A quick summary of identified problems:

  "Chief, we have a fire and rescue problem on the second floor, and a large smoke and rescue problem on the third floor above the fire."

- Current resource assignments – A brief summary of what companies are assigned to specific areas and functions on the incident:

  "So far, we have Engine 1 attacking the fire on the second floor, I have given my company (Engine 2) away to the captain on Engine 1 as the initial 'two out' for rapid intervention, and Truck 1 is heading to the roof to open up the penthouse door over the stairwell."

- Current IC's organization – An explanation as to what ICS "tools" have been utilized thus far:

  "I have designated Engine 1's captain as Division 2 Supervisor on the second floor and Truck 1's captain as the Ventilation Group Supervisor."

- Non-resourced priorities – A brief summary of tactical objectives that have not been assigned for companies to accomplish:

  "Chief, the rest of the alarm assignment is in Level 1 staging down the street, and I have not yet addressed the rescue problems on the second and third floors or the salvage problem on the first floor."

- New assignment for the initial IC being relieved – An important question to be asked and one that is often overlooked. The officer that is relieved should never walk away from this face-to-face transfer of command without a clear understanding of what his or her role on the fireground is, once relieved of incident command responsibilities.

As the above format is followed, the right quantity of information will be shared, and a pertinent dialogue will be initiated. As mentioned earlier in this module, the use of a command worksheet is a valuable tool for an effective transfer of command. Proper documentation on this worksheet will enable the initial incident commander to recall and share the above command transfer components quickly. One of the most appreciated things that a chief officer can receive is a well-organized incident that is following prudent strategy and a subsequent set of tactical objectives. Beyond that, receiving a clear and concise command worksheet from the initial incident commander ensures that the continuity of command will not be broken during crucial moments when fire incidents can be the most volatile and demanding.

THE USE OF A COMMAND WORKSHEET IS A VALUABLE TOOL FOR AN EFFECTIVE TRANSFER OF COMMAND.

MODULE

Fireground Risk Management 3

Objectives

After completing the reading and activities in this chapter, students should be able to:

OBJECTIVES

- Identify common factors contributing to firefighter injuries and fatalities.
- Identify the function and importance of a fireground accountability system.
- Execute federal and state regulations regarding "two-in/two-out" requirements for environments that are immediately dangerous to life or health.
- Understand and execute the functions of a rapid intervention team.
- Identify the importance of monitoring structural environments during post-fire overhaul operations.
- Understand command and control concepts, utilization of incident management systems, and their relationship to reducing risk on the fireground.

Factors Contributing to Firefighter Injury and Death

FIREFIGHTING STATISTICS

Firefighting is one of the most dangerous occupations known to man. As we look at firefighting fatalities around the globe, we see that the United States experiences more fatalities than any other country. Is it more hazardous to fight fire in the United States? That question will evoke debate from many factions of the fire service, but here are the facts: Many other countries require more training for fire officers and more education for incident commanders than is required in the U.S. In addition, other countries have a more stringent system of accounting for their personnel and for each firefighter's air supply than does the typical North American fire department.

We must not be afraid to look at how firefighters have been injured and killed at fires. As retired Deputy Chief Vincent Dunn of the New York Fire Department has observed, "In order for a firefighter to survive the dangers of firefighting, he must know how other firefighters have died or been seriously injured."

LEARNING FROM OTHERS' EXPERIENCE

There are numerous activities available which reinforce lessons learned from others' experience. Among these are holding post-incident reviews, analyzing news tapes, and listening to audiotapes of fires. Company officers should have a regular process of refreshing themselves with information from books, journals, Internet websites, and other sources of case studies. A mark of a true professional is continuous self-improvement and a mind open to what others have done.

By looking back, we can identify some common factors that have contributed to injuries and death among firefighters. Factors that are identified in numerous reports include the following:

- communications failure
- no SOPs/SOGs or not following SOPs/SOGs
- lack of an effective incident command system
- lack of an effective fireground accountability system

In this module, we will look at these four contributing factors. We will also examine safe operating practices for the fireground.

CAUSES OF FIREFIGHTER INJURIES AND FATALITIES

What are some of the common causes of firefighter injuries?

- strains and sprains
- falls
- overexertion
- smoke inhalation
- exposure to communicable diseases
- responding/returning

What about causes of firefighter fatalities?

- building collapse
- being struck by a falling object
- heart attack/stroke
- becoming trapped/lost/disoriented

WAYS TO PREVENT INJURIES AND FATALITIES

As individuals, we can prevent many injuries by keeping physically fit, practicing good nutrition, and engaging in a regular stretching program. In addition, we can avoid many dangerous situations by establishing collapse zones, performing solid risk/benefit analyses, not committing firefighters to "no win" situations with questionable benefit, and utilizing a fireground accountability system with adequate staffing for working incidents.

Communications Failure

Communications problems are not just about hardware. Certainly there are radio systems that are plagued with dead spots, busy signals, and other signal problems, but such systems do not account for the majority of fireground incidents. Poor management of the communications process by incident commanders is probably the main infraction.

In several tragic incidents, trapped firefighters were heard transmitting their location and situation to command staff who weren't even listening. Other incidents have been documented in which the radio discipline was so poor and the process was so gravely mismanaged that the units with critical emergency messages were not able to get their situation to command or dispatch.

PPN method

REPORTING POSITION, PROGRESS, NEEDS

One method of fireground communication that helps to solve some of these issues is known as the "PPN method." PPN is an acronym that stands for Position-Progress-Needs. With every transmission, fire companies restate their position, their progress toward completion of their current assignment, and any needs, including additional resources, repositioning of adjacent forces, ventilation, lighting, etc. An example of this process follows. In this scenario, Engine 2 is assigned to conduct salvage on the first floor. The fire is being attacked on the second floor.

- After receiving assignment:

  "Command, Engine 2 has been assigned as the Salvage Group. We are setting up to work on the first floor with six personnel from Engine 2 and Ladder 1."

- Progress report:

  "Command, Salvage Group. We are actively completing salvage operations on the first floor. We're approximately 50% complete. We're having a problem with runoff in the offices on the southeast corner. Could you have a company in Division 2 try to reroute some of that flow?"

- Upon completion:

  "Command, Salvage Group has completed the salvage operations on the first floor. All personnel are accounted for. We're ready for reassignment. We do not require rehab."

The above process provides a framework for radio communication and eliminates the wasted airtime associated with a free-form communication style.

Benchmarks

MEASURING PROGRESS TOWARDS CONTROL OBJECTIVES

Benchmarks are measuring points or milestones transmitted by fire companies to command. These "tactical benchmarks" give the command post a quantifiable sense of progress toward the control objectives. Upon a company's arrival at an incident, the first benchmark is, "Primary search complete." The second is, "All clear." After receiving these two benchmarks, the incident commander knows that the focus of the operation can now move from life safety/search and rescue to fire control. The next fire control benchmark is, "Fire under control," followed by, "Loss stopped."

As a large area search progresses, an "all clear" is transmitted from each floor as the search is completed in that area. In this way, the commander gets a feel for the amount of time the search is taking and the resources required. When long periods of time elapse between benchmarks, no progress is being made. This means that either the plan is flawed or more resources are needed.

Communication aids

USING REMOTE MICROPHONES

Remote microphones can assist all firefighters in hearing critical messages and details. They also allow more information to get back to the command staff when company officers don't have to stop, take off their gloves, and retrieve a portable radio from a hip pocket or a turnout coat pocket. If the mic is handy, officers will use it!

Making assignments clear and direct gives company officers a focus and a goal. "Focus" is nothing more than going in with a purpose. As companies are assigned, have them trade their company designator (Engine 2, Ladder 1) for a tactical designator (Roof Division, Rescue Sector, Attack Team 1). This designation makes it harder for the crews to free-lance and assign themselves to another function without authority. The "Roof Division" shouldn't be extending an attack line on the third floor!

USING TACTICAL DESIGNATIONS

Another factor that assists the communication process is having company officers "act like officers." Standard practice should require officers to manage and supervise. These roles require officers to direct, listen, speak, and focus. Firefighters are the workers who complete the hands-on tasks.

OFFICER CONDUCT

No SOPs/SOGs or Not Following Them

It is difficult to believe that in the litigious society in which we live, many fire departments operate without standard operating procedures (SOPs) or standard operating guidelines (SOGs). In these departments, the officers make up the rules or the game plan on the fly. Every event is an exercise in "management by crisis." Without policies and guidelines, there is no development or enforcement of standards. Firefighter safety is one of the areas that is adversely affected. In other departments, policies and guidelines exist, but there is no training or formal process to put what is in the book into practice in the real world. When there is no training and no enforcement, safety violations will continue to occur and increase.

Examples of SOPs/SOGs that affect firefighter safety and survival include the following:

SOPs/SOGs AFFECTING FIREFIGHTER SAFETY AND SURVIVAL

- written risk policy
- fireground accountability
- evacuation signals
- free-lancing

If we ignore actions on the fireground (or elsewhere) that are unsafe or contrary to procedure, then we accept (and condone) such actions. Chief Alan Brunacini of the Phoenix, Arizona Fire Department has stated, "If an SOP is not being followed, it should be changed so that it can be enforced, eliminated if it can't be enforced, or enforced."

Lack of an Effective Incident Command System

There are three basic incident management systems present in North America:

- Incident Command System (ICS)
- Fireground Command System (FGC)
- Incident Management System (IMS)

The Incident Command System is a system that was derived from concepts that came out of the California FIRESCOPE program. The system was created after some devastating wildland campaign fires in the early 1970s. It has been adopted by the National Fire Academy as the preferred incident command system in all courses and curriculum. It has also been adopted for use by all federal wildland agencies as well as the U.S. Coast Guard for managing hazardous materials incidents.

INCIDENT COMMAND SYSTEM

FIREGROUND COMMAND SYSTEM

The Fireground Command System was developed in Phoenix, Arizona, for structural fire attack. The system has been made popular both nationally and internationally by Phoenix Fire Chief Alan Brunacini in his book, *Fire Command*. The system has been continually refined to include all-risk and all-hazard emergencies.

INCIDENT MANAGEMENT SYSTEM

The National Incident Management System Consortium proposed merging the ICS and the FGC into one national Incident Management System. The consortium developed model guides for various types of emergencies, but the IMS they created did not succeed in merging the other two systems. Rather, it actually created a third system that many agencies have adopted as their primary emergency management system.

WAYS TO IMPROVE EFFECTIVENESS

The use of any of these three systems is permitted and encouraged. All three systems are based on unity of command, span of control, and management by objectives. Regardless of which system your agency has chosen, there are a few considerations for improving the effectiveness of any system. First of all, agencies must use the system on every incident. If only large-scale emergencies are managed with the system, personnel will not be proficient enough to work effectively. Secondly, the Incident Command System is built on "focus." Each objective and task assignment should be completed quickly and efficiently, without interruption or distraction. Single assignments to fire companies based on priorities will keep each company focused on results.

Personnel should be trained and experienced in every ICS position to which they are assigned. Having a department-wide program, in which all personnel must complete didactic training and then perform the position duties under supervision on a real or simulated full-scale exercise, will provide the necessary education and experience for each member to function effectively as an incident manager. Position taskbooks which must be signed off before a member can become certified in every ICS position, such as those published by the National Wildfire Coordination Group, may prove helpful in documenting training.

Lack of an Effective Personnel Accountability System

Accountability is both a system and a process. It is a relatively simple concept that seems to have been made difficult by our bureaucratic structure and affiliated organizations. Fireground accountability is both an attitude and a commitment:

I will take care of myself, and I will take care of others on my team.

"I will take care of myself, and I will take care of others on my team."

The concept is that every individual will commit to the following five behaviors:

1. I will follow the rules, the SOPs, orders, and good common sense.
2. I will not become a "rescue" myself by being too aggressive or too complacent.
3. I will not free-lance, nor will I allow others on my crew to free-lance.
4. I will check out and wear my "stuff" properly and use all job aids.
5. I will communicate my position, progress, and needs to my supervisor with every radio communication.

Under a personnel accountability system, every firefighter is responsible for his/her own safety and accountability. Every firefighter is also responsible for his/her partner's safety and accountability. Every supervisor is responsible for the members assigned to his/her unit, company, sector, division, or group. If these standards are followed, then most firefighter fatalities would be prevented. Many national standards and department procedures require the use of such a fireground accountability system.

Using a Fireground Accountability System

There are numerous accountability systems on the market. Among them are magnetic boards, velcro tags, passports, helmet tags, ID cards, radio controlled PASS devices, computer based tracking systems, and pen-and-paper riding lists. There are pros and cons with every system, but there are a few common features that every system should possess. A good accountability system should have the following components:

COMPONENTS OF AN ACCOUNTABILITY SYSTEM

- It is cost effective.
- It has a backup.
- It can be easily taught, practiced, and remembered.
- It can be used on both large and small incidents.
- It can be integrated with mutual aid incidents.
- It is used on every incident.
- It requires fire companies to stay together. Crew integrity is the primary operational factor!

The purpose of a fireground accountability system is to ensure that all personnel operating on an emergency are accounted for.

The purpose of a fireground accountability system is to ensure that all personnel operating on an emergency are accounted for. There have been many instances of firefighters being found missing after a collapse or flashover, with no record of their last whereabouts. In Texas just a few years ago, a firefighter who responded to a working fire drove directly to the scene and was trapped in a collapse. Rescue crews did not even know that he was operating in the building, since he had followed an attack line into the structure and had "assigned himself" to an interior crew, without checking in with the incident command post or staging location.

We must ensure that all personnel are working in teams, that they are performing valid assignments given by the incident commander, and that they are working in concert with all other assigned units and companies. An accountability system provides a reference of who is working and who is responsible for each person's supervision. A breakdown often becomes evident when a firefighter is reported missing or trapped, and the "system" does not know the last location where the firefighter was working. In Worcester, Massachusetts, in December 1999, firefighters operating inside a vacant cold-storage warehouse frequently changed work locations and floor levels without transmitting each and every change to an outside party. Using tactical designators on the radio, such as "Division 3" for the third floor and "Division 4" for the fourth floor, in place of "Rescue 1" or "Engine 5," might have assisted rescue crews in determining the last known position of trapped or missing personnel.

Although the IC is responsible for assuring that all crews are accounted for, the accountability system frequently resides under the Planning Section of the Incident Command System. The resources status unit or "restat" will normally manage the accountability process. Accountability is required at all incidents. It is the process whereby the general location and function of crews are tracked. The specific location of crews within a structure is not tracked. The goal of an accountability system should be to track the location of all crews and units at a fire or other emergency, both inside and out!

Accountability is the process whereby the general location and function of crews are tracked.

The key to transporting all of your personnel back home after the emergency, in the same condition in which they arrived, relies on the decisions made by the company officers and incident commanders during fire attack. Are you allowing your people to free-lance? Do they set their own tasks, assignments, and positions? Are you aware of every member's whereabouts and assigned task at every minute of the operation?

A "sudden hazardous event" requires answers to two questions:

- Are crews in the affected area?
- If so, who are they, and how many members are we looking for?

The accountability system must provide the location of all crews, regardless of whether they are assigned at the incident or preassigned based on order of arrival. In either case, incident-specific assignments should be used to track crews from the moment of assignment. When preincident assignments are used, companies must transmit their "PPN" (position, progress, needs) as soon as possible.

PERSONNEL ACCOUNTABILITY REPORTS

Another phase during which firefighters may become lost or unaccounted for is during crew rotations. All crew rotations, including rehab, SCBA bottle change, and floor/area reassignment, require a "PAR" (personnel accountability report). PAR is the process whereby an officer reports that he/she has accounted for the whereabouts and safety of every assigned member. There are two types of PARs:

- Emergency PARs – Done at the call of the IC or at evacuation.
- Routine PARs – Done whenever the crew leaves the structure, until the fire is "under control."

One may think that firefighters get lost only on large, complex fires in large buildings. On the contrary, firefighters are frequently lost or trapped inside one- and two-family homes. In 1995, three firefighters became lost at the Bricelyn Street Fire in Pittsburgh, Pennsylvania. They were missing for 1 hour and 20 minutes before their absence was noticed. Routine PARs would have helped in this situation.

MULTIPLE ACCOUNTABILITY ENTRY POINTS

At large incidents or those of longer duration, there are more frequent crew changes, with more opportunities for personnel to become lost, disoriented, trapped, or unaccounted for. One aid in keeping track of personnel in such situations is multiple accountability entry points. Such a setup allows firefighters to enter and exit from multiple sides of an incident.

Two-In/Two-Out Rule

WHAT THE RULE SAYS

In January 1998, the Occupational Safety and Health Administration (OSHA) passed the "two-in/two-out" rule. This rule has since been included in a number of model standards and organizational SOPs. The concept is simple. Upon initial arrival at a fire or rescue incident where the environment is immediately dangerous to life or health (i.e., a working fire), the initial entry team shall be no less than a two-person team. This entry team may not make entry prior to the arrival or assembly of an additional two-person team that will remain outside the structure in a ready posture to perform rescue of the initial team, should it become necessary.

It is essential that a two-person team remain outside the hazardous area, in the event the initial team requires assistance.

The practice of a two- or three-person engine company arriving at a working fire and entering the hazardous atmosphere to perform firefighting, search, or rescue, with no backup on the scene ready to provide immediate rescue, is unsafe and should not be common practice. It is essential that another two-person team remain outside the hazardous area, in the event the initial team requires assistance.

The incident commander has the flexibility to determine whether more than two outside firefighters are necessary when more than two firefighters go inside. In a situation in which the burning structure is very large, additional outside firefighters may be warranted to ensure effective assistance and rescue. For example, where the firefighting involves entry from different locations or levels, two outside firefighters may have to be stationed at each point of entry.

EXEMPTIONS

There is an explicit exemption in OSHA's Respiratory Protection Standard that if life is in jeopardy, the two-in/two-out requirement is waived. The incident commander and the firefighters at the scene must decide whether the risks posed by entering an interior structural fire prior to the assembly of at least four firefighters are outweighed by the need to rescue victims who are at risk of death or serious physical harm. There is no violation of the standard under rescue circumstances.

INTENT OF THE RULE

The intent of the two-in/two-out rule is a worker safety practice requirement, not a staffing requirement. The standard does not require fire departments to hire additional firefighters, nor does it require four-person fire companies. Most fire departments have more than four firefighters and can assemble the numbers required on the scene by waiting for others to arrive. During this time, the fire may be attacked from the outside, size-up operations may occur, and emergency rescue necessary to save lives may take place. Additional staff can be assembled by such means as calling for a second fire company at the scene, calling in additional firefighters who are on standby, and calling on other nearby fire departments when necessary.

This OSHA standard, like all other federal OSHA standards, only applies to public and municipal workers such as firefighters in states which operate their own occupational safety and health program under an OSHA-approved State Plan. Any state with an approved State Plan must have standards which are "at least as effective as" the federal standards and must extend their coverage to state and local government employees. There are other national model standards and state statutes that may also apply. Check with the authorities in your own jurisdiction regarding local requirements.

Rapid Intervention Teams

Another idea to support and add an extra margin of safety is the rapid intervention team or RIT. The mission of rapid intervention is to search for and remove lost, trapped, or injured firefighters. Primary responsibilities of the RIT will include but not be limited to the following:

> The mission of rapid intervention is to search for and remove lost, trapped, or injured firefighters.

- searching the area where the missing/injured firefighters were last reported
- following the hoseline of lost firefighters to minimize rescue time
- looking and listening for evidence of lost firefighters
- searching hallways, exterior walls, and collapsed areas
- monitoring radio transmissions of operating companies
- reporting all findings and progress to the incident commander

The RIT will report to the incident commander or command post with the following tools and equipment (these are the minimums):

- full turnout gear including SCBA
- forcible entry tools (flathead axe, halligan tool, sledgehammer or maul, short pike pole, saw, etc.)
- portable radios, handlights, search rope (150 feet minimum)

The purpose of the RIT is to have a crew all dressed up and ready to go if a crew or an individual firefighter gets in trouble. Once in place, the RIT should have no other focus than rapid intervention activities!

WHEN A RAPID INTERVENTION TEAM IS REQUIRED

When is a rapid intervention team required? NFPA 1500 states, "A team is necessary any time crews could be in immediate danger." At a minimum, a RIT should be assigned at all commercial structures and at the call of the incident commander. A RIT should also be assigned on other complex and difficult fires where firefighters are historically required to "extend" themselves to attempt to save life. Some of these include:

- multifamily occupancies
- commercial buildings
- shipboard fires
- high-rise fires

Fire companies assigned to rapid intervention duty should receive specialized training, including the following basic requirements:

- saving our own
- rope rescue techniques
- advanced SCBA
- confined space training

The mission of the rapid intervention team must remain clear to all personnel, but especially to chief officers. That mission is: *We only act to save the life of a fellow firefighter*. If Command needs more crews for other functions, then Command needs to order more crews! It is a temptation to grab the RIT to extend one more hoseline, rather than to wait for a company from staging or from quarters. Commanders must resist this temptation. Hands off the "safety net"! Once the RIT has been assigned, there is no immediately deployable team if something tragic occurs.

Some of the tools that have proven valuable for RIT usage are the same ones that rescue companies and truck companies use for extrication and structural collapse. Each team should select tools specifically for the type of building construction and the materials present at each fire. A typical list of tools is presented in Figure 3-1.

Typical Tools Used by a Rapid Intervention Team

Lifting tools
- Maxi Force® air bags
- hydraulic spreaders/rams
- Paratech® air shores

Supports
- Paratech® long shores
- Paratech® raker shores

Breaching/breaking tools
- halligan/irons
- packhammers
- Cutter's Edge® concrete saw
- Cutter's Edge® chainsaw

Breathing supplies
- air carts
- SCBA/SABA
- spare cylinders

Tools for moving the injured
- stokes basket
- scoop stretcher

Visibility tools
- Bullard® thermal imaging camera

Miscellaneous
- door chocks and handlights
- rescue rope and necessary tools
- hauling line

Figure 3-1

STAGING FOR RITs

On a typical low-rise structure fire, the rapid intervention team will routinely stage at the command post. This allows the team to be ordered verbally into service by the incident commander, without the fear of a communication delay or glitch. The RIT should familiarize themselves with the accountability/resources status board and make note of where crews are working. On incidents where the possibility exists for crews to get into trouble remote from the command post, a second RIT may be assigned. At high-rise fires, the RIT should position themselves at staging or base.

RIT SCENE ASSESSMENT

A company being assigned RIT duties must conduct a separate scene survey. The purpose of this assessment is to gain knowledge of the incident and the structure so as to make a proper selection of tools and techniques needed for the job. The size-up starts with a 360-degree walk around the structure. During the size-up, the team should be alert for the following types of information:

- construction type
- quick ways in and out
- fire escapes/ladders/balconies
- roof type
- ledges
- window configuration and type

On most fires, the RIT will not be activated. The temptation to under-prepare and not be diligent must be resisted. This is an important assignment that must be handled with confidence and faithfulness. When the RIT does get activated, it will be a moment unlike any other in your career. There will be more pressure and conflict than you might imagine. Seasoned veterans have been known to scream and panic, while average firefighters have been observed performing heroic acts. The company officer assigned to the RIT crew must maintain discipline, order, and a calm demeanor throughout the RIT activities. Panic breeds panic. Firefighters are not used to being in the middle of their own emergency. "Managing the mayday" and maintaining control of your forces and equipment will be a formidable task.

MANAGING A "FIREFIGHTER DOWN" REPORT

A "mayday" or "firefighter down" report must be managed properly. The incident commander puts out a tone requiring all companies to clear the air and maintain radio silence. Command then transmits directly to the unit with the emergency traffic. Plain English and clear text should be used. The last location known should be requested, and Command should assign the RIT and any available companies in staging to RIT support. Fire companies actively involved in fire attack and support activities should not be allowed to reassign themselves to the rescue. Command must confirm the assignments for all companies and specifically designate those companies assigned to the rescue. The rescue should be moved to a separate frequency, and companies not directly assigned should stay off that channel.

After the initial broadcast and RIT assignment, Command must evacuate all non-essential crews and conduct PAR. This step is critical in "corraling" firefighters that may self-assign themselves to the rescue and may become lost or trapped. It also provides the opportunity to rehab a large number of firefighters for the next wave of search, rescue, or other RIT support functions.

It is critical to continue fire control objectives throughout the mayday.

It is critical to continue fire control objectives throughout the mayday. Safety officers must be aware to watch for signs of structural collapse and other hazards that may be missed as firefighters "tunnel vision in" on the rescue of one of their own.

Monitoring the Structural Environment

POST-FIRE CONCERNS

Many fireground injuries occur after the fire attack has been completed. Injuries during salvage and overhaul are very common. Some of the factors involved include the weakened condition of the structure, static water loads causing floor/ceiling collapse, post-exertional fatigue, and cardiovascular stress. The inhalation of carbon monoxide after SCBA is removed is another common but preventable concern. In the post-fire environment, carbon monoxide levels are increased due to incomplete combustion resulting from the interruption of the fire tetrahedron. As smoldering materials continue to produce smoke, toxic levels of fire gases are produced.

MONITORING CARBON MONOXIDE LEVELS

Assessment of the post-fire environment should be made to ensure that structural integrity, trip and fall hazards, weakened floors, and depleted oxygen levels have been addressed. Mechanical ventilation and portable lighting can be utilized to improve the conditions in the working area. SCBA should be mandatory until air monitoring has confirmed that the carbon monoxide level is within acceptable limits recommended by your department or the state occupational health agency.

Reducing Risks

Risks are probabilities of occurrences. The risk of death or injury is present in most activities. Hazards are concrete situations. You can identify hazards. When you take action to control, reduce, or even eliminate a hazard, we call that risk management. For example, when the temperature is below 32°F (0°C), water will freeze, and walking surfaces will become slick and slippery. Having the driver/operator or engineer sand or salt areas around fire apparatus in cold weather is an example of a hazard and a mitigating action. Another example would be crews operating on a significant fire in an unreinforced masonry building. In this situation, one of the hazards is structural collapse. The mitigating action would be to establish a collapse zone 1-1/2 times the height of the building and to have all apparatus and personnel remove themselves from potential harm.

EXAMPLES OF HAZARDS AND MITIGATING ACTIONS

Along with the size-up and the initial tactical orders, the initial incident commander must perform a risk analysis. Risk analysis is a mental evaluation in which the risks of the current situation are weighed and compared against the benefits of a successful outcome (Figure 3-2).

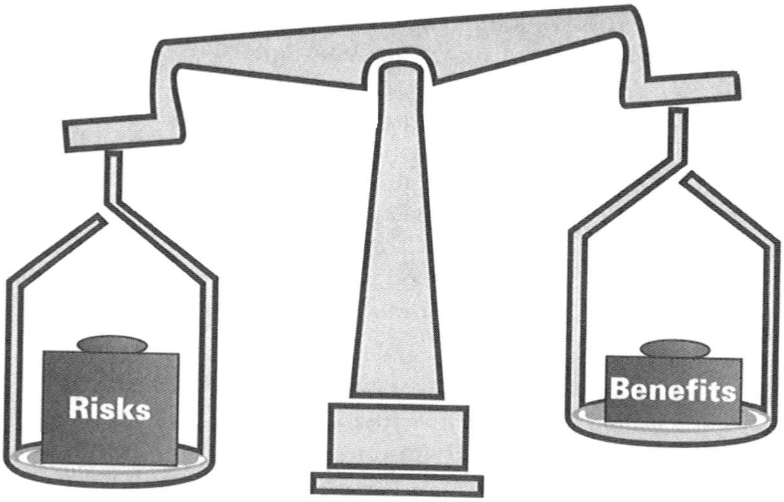

Figure 3-2: Risks must be weighed against benefits.

EXAMPLES OF RISK ANALYSIS

For example, you may assess a vacant, commercial structure that is 70% involved as too hazardous for interior operations. The risks are possible roof collapse, floor collapse, firefighters becoming lost/trapped, etc. The benefits are that the building is vacant, there is no property value, and the building will be torn down after the fire. In this situation, the risks are far greater than the benefits of committing vast amounts of resources and trying to ensure their safety. On the other hand, for a well-involved fire of two rooms on the first floor of a residential structure, with a confirmed occupied second floor, the risks are great, but the benefits are even greater.

RISK VS. GAMBLE

What is the difference between a risk and a gamble? It's the same definition used in Russian Roulette at a Las Vegas casino. A risk is a situation in which you have at least a 50/50 chance of being successful. A gamble is one in which your odds of winning are less than 50%. Remember, this job is tough enough. We will take risks, but we don't gamble with our personnel!

Classic Risk Management

COMPONENTS OF CLASSIC RISK MANAGEMENT

Company officers should understand the basics of risk management, in order to have a better understanding of their role in making the job safer. Classic risk management is a process used by every industry to identify and manage risk. It is divided into five steps:

1. **Risk identification.** This is the process of assessing all the things that might go wrong at an emergency and developing a "worst-case scenario" list. At a working fire, the Planning Section will typically generate a forecast of what threats are present or may potentially be present. This process is risk identification in action.

2. **Risk evaluation.** The second step involves evaluating each item on the above list from the standpoint of frequency (how likely it is to occur) and severity (how much damage can be produced by the risk). This frequency and severity assessment can be seen in the risk matrix in Figure 3-3.

Risk Matrix	
I. High Frequency Low Severity	III. High Frequency High Severity
II. Low Frequency Low Severity	IV. Low Frequency High Severity

Figure 3-3

3. **Establishing priorities.** The third step is establishing priorities. Incident managers and safety officers must address the items with the highest frequency and highest severity first. In the risk matrix in Figure 3-3, risks that are categorized in Quadrant I or II (low severity with either high or low frequency) are normally the lowest priority. Quadrant III risks (high frequency/high severity) are much more serious and include events such as flashover, being struck by falling objects, and the like. Most firefighters are well aware of these risks. Quadrant IV risks (low frequency/high severity) are often overlooked because of their low frequency. Establishing risk priorities on each incident will identify unique hazards and threats.

4. **Risk control.** Risk control is the step in which mitigating measures are discussed and agreed upon. It is the process of developing methods to eliminate, reduce, or control each risk. Requiring the use of SCBA and full personal protective clothing will eliminate many Quadrant III and IV risks. A daily stretching and physical fitness program can reduce or eliminate many Quadrant I strains and sprains.

5. **Risk monitoring.** The final step in risk management is the enforcement and measurement of the risk control activities. Peter Drucker, a renowned management expert, has stated, "If you don't measure it, you don't manage it." Enforcement and evaluation of the risk control measures that have been applied will glean much information as to the effectiveness of the overall risk management program.

Summary

Risk management is the process by which fire officers assess, eliminate, or control hazards on the fireground. Certain functions and activities have been shown to improve the safety and survival of firefighters at structural incidents. Some of these include adequate staffing, rapid intervention teams, personnel accountability systems, and sound decision making by experienced and educated fire officers. The National Fire Academy's Risk Policy condenses this module intent very well. It says:

NATIONAL FIRE ACADEMY RISK POLICY

"We will take great risk to save life.
We will take some risk to save property.
We will risk nothing to save what is already lost."

MODULE

4

Tactical Deployment of Engine and Truck Company Operations

Objectives

OBJECTIVES

After completing the reading and activities in this chapter, students should be able to:

- Identify the five basic functions of engine company operations.
- Identify the 10 basic functions of truck company operations.
- Explain the value of predesignated "seat assignments" in relation to fireground engine company and truck company responsibilities.
- Execute the tactical deployment of an initial alarm assignment responding to "residential and commercial" structure fires.

Engine Company Operations

The engine company is the workhorse of fire suppression efforts. Every fireground function that exists in modern firefighting exists to support the most critical operation: putting water on the fire. We sometimes minimize this activity. Contemporary textbooks discuss ventilation, forcible entry, laddering the building, positioning aerials, search, rescue, and a myriad of other operations, but the bottom line is that if trained and equipped firefighters can't get an efficient fire stream into operation at the seat of the fire, extinguishment is not going to happen soon enough for any of us.

Role of the Engine Company

The role of the engine company is to provide fire attack activities through the selection, extension, and operation of firefighting water and extinguishing agents. This role is obviously based on function. The triple-combination pumper has three components: water, hose, and a pump. These components correlate with the three sides of what is known as the "water triangle": supply, pressure, and power (Figure 4-1).

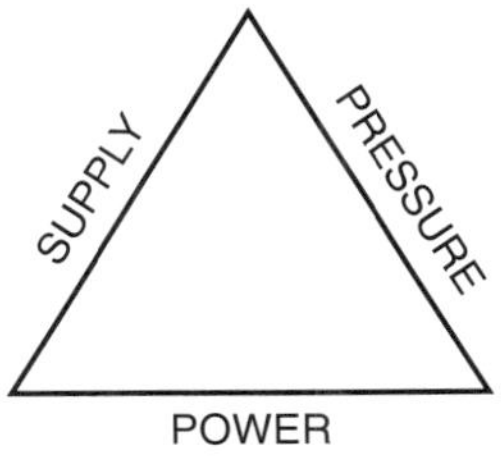

Figure 4-1: The water triangle.

A few assumptions regarding the water triangle must be made:

ASSUMPTIONS REGARDING THE WATER TRIANGLE

- Maximum flows are desired; therefore, the water delivery system will be expected to perform to that requirement.
- The system must be flexible, because of the dynamic nature of fire.
- To achieve maximum flows, the pumper must be taken to one of the three limits.

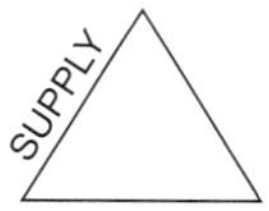

Water supply

The biggest single problem in most major pumping operations is water supply. This area has the greatest effect on the total water flow available. Usually, improving the water supply requires more effort than dealing with the other limiting factors in the water triangle, but the results are always worth it. The limitation of this side of the triangle is indicated by the compound gauge showing vacuum. Some ways to improve the water supply are as follows:

WAYS TO IMPROVE THE WATER SUPPLY

- Utilize additional hydrants.
- Drop or lay additional supply lines.
- Choose a large-diameter hose in place of smaller supply lines.
- Use portable tanks.
- Initiate a tanker/tender shuttle.

Any combination of these factors can also assist in improving the water supply side of the water triangle.

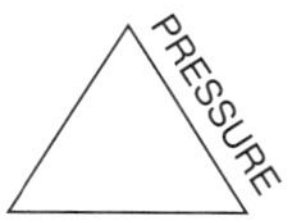

Pressure

The second leg of the water triangle is pressure. This is another limiting factor in developing maximum flows. The most common reason for pressure problems is using hoselines that are too small in diameter. With today's technology, this factor is the easiest to improve. Each type of hose, appliance, or equipment affects pressure differently. It is important to conduct flow tests to determine exactly which components are stealing pressure. The limitation of pressure is usually indicated by the master pressure gauge.

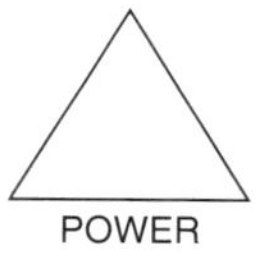

Power

The final leg of the water triangle is power, meaning the power available from the pumping engine. With today's fire apparatus, this is rarely a concern. The average pumper in the United States is ordered with a 300-plus HP turbodiesel engine. Lack of power is the last limitation in a major water flow operation. This limitation is indicated when the throttle is at maximum and the engine or pump is also at maximum output. The system can be improved to perform better by bringing additional pumpers online, using larger, more efficient hoselines, and upgrading rigs through the budget process with larger capacities. With apparatus that is poorly maintained or not regularly pump tested, power limitation can be a problem.

Fire Attack Options

DIRECT ATTACK

There are two basic fire attack methods: direct and indirect. Direct attack is pretty easy to understand: "Put the wet stuff on the red stuff!" Actions aimed at locating, confining, and extinguishing the fire are the activities of choice. The number of options that these activities entail could be thousands, depending on the occupancy, construction, number of floors, access, proximity to a fire suppression water supply, and proximity to building suppression systems.

Indirect attack involves protecting exposures, which includes the fire attack activities of blocking, directing, and extinguishing remote portions or fingers of the fire until an attack can be made on the head or seat of the fire.

INDIRECT ATTACK

It is the responsibility of the engine company officer to be an expert in his or her district and to have a working knowledge of the building stock, use, and construction of the fire building. The engine company officer must also be an expert on the water system capabilities and limitations, including the main size, grid configuration, and hydrant locations.

Primary Engine Company Functions

There are five basic engine company functions:

1. Locate.
2. Confine.
3. Extinguish.
4. Overhaul.
5. Secure the scene.

Locating the Fire

One of the first priorities for the first arriving engine company is locating the fire. Determining the exact location will set the stage for the actual tactics and tasks that must be performed and will better equip the command staff as to the true extent of the problem. If a fire is showing from a first-floor window and is auto-exposing the floor above, the approach will be much different from the approach used if the fire is confined to a small trashcan on the second floor with a widespread smoke condition. The location of the fire in regard to exits, windows, vertical stairwells, ventilation, and elevator shafts is also significant.

HOW FIRE LOCATION AFFECTS TACTICS AND TASKS

Many fires gain incredible headway while engine company personnel are looking for the source. In large buildings where smoke is all that is visible, a rapid, systematic search for the fire must be initiated. Using their knowledge of fire behavior, fire officers can look at common avenues of fire travel and smoke spread. The more quickly the fire is located, the faster the fire attack can commence. Fire loss has a direct correlation to the speed with which the main body of fire is located. Thermal imaging cameras can dramatically improve the search for hidden fire in attic spaces, walls, floor assemblies, HVAC system components, and other concealed locations.

Fire loss has a direct correlation to the speed with which the main body of fire is located.

Confining the Fire

Confining the fire is another major function of engine companies. The importance of confinement relates to the life hazard present in the building and the amount and location of available, unburned fuel. The first line should normally be placed between the fire and any trapped occupants. This may require more than one line in a large area or fire that is progressing rapidly. In center hallway or multistory apartments and hotels, oftentimes the first line will be brought into the interior hallway, while the fire attack stream is advanced through an outside balcony, slider, or fire escape to extinguish the fire. The hallway line provides confinement and holds the fire to the room of origin.

The first line should normally be placed between the fire and any trapped occupants.

Coordinating multiple fire streams to complete the same objective is often difficult. It is common to have some communication problems on most working structural fires. A quick, sustained fire attack at the head or main body of fire can slow the spread of fire and improve confinement. Confinement is much more

GETTING THE FIRE UNDER CONTROL

difficult on large fires, where the seat of the fire has not been reached. In these cases, multiple lines must be operated on remote portions of the fire until the attack teams can advance to the seat of the fire. Selecting an inside/out attack often improves confinement as the fire is pushed outside the structure. In sealed buildings or in those cases in which an outside/in attack is chosen, additional engine companies should be considered to control the fire's travel and effect confinement. The benchmark transmitted when the fire has been confined is, "Fire under control."

Extinguishing the Fire

TASKS INVOLVED IN EXTINGUISHING THE FIRE

Extinguishing the fire is the broad goal of all engine company activities. The tasks involved include selecting the proper hoseline, selecting the proper stream, advancing the stream and team to the fire, controlling fire spread, and providing an adequate volume of water to cool and suppress the fire. Statistics show that 90% of all structural fires are extinguished with two lines. The remaining 10% of the fires require multiple lines and account for the vast majority of fire loss in the United States. On deep seated fires, the extinguishment activity may take several hours of flowing water.

Overhauling the Area

STOPPING DAMAGE

Overhaul is the activity wherein all the burned area is checked for smoldering embers. During this process, the structure is checked, concealed spaces are opened up, and the structure and burned contents are thoroughly soaked with water. All unburned and burned contents are separated, and the potential of a new fire rekindling in the structure is eliminated. When all fireground activities are complete and there is no further damage from the fire or from water streams, the benchmark that is transmitted on the radio is, "Loss stopped."

COORDINATING OVERHAUL WITH FIRE INVESTIGATION

The overhaul process must be coordinated with the fire investigation. All fires must be investigated for cause and origin. Some fires may be investigated for arson or other criminal activity. A complete overhaul frequently destroys evidence and pattern that the fire investigator will need to identify the cause and point of origin. Engine company personnel should not start the overhaul process without an order from the incident commander, after consulting with the assigned investigator.

Securing the Scene

INSURING THAT THE PROPERTY IS LEFT IN A SAFE CONDITION IS OF PARAMOUNT IMPORTANCE.

In many systems, the first-in engine ends up being the last company relieved. The responsibility for securing the scene normally rests with the last company officer leaving the scene. Insuring that the property is left in a safe condition is of paramount importance. Fire departments have been liable for injuries sustained by the public after leaving a fire building unsecured. Posting caution tape around the scene, shutting off utilities, boarding up exits and windows, and eliminating hazards such as burned-out stairways can help make the scene safer for those that must perform other services in the days ahead. Many departments will request heavy equipment from the public works department to knock down unstable walls or roof members to prevent a future collapse injury.

Safety During Interior Operations

Interior fire attack is hazardous. There are three hazardous events that continue to endanger firefighters. These are flashover, structural collapse, and firefighters becoming lost or trapped.

Flashover

Flashover is an event that can kill firefighters. This phenomenon predictably occurs in the middle of the free-burning stage of a fire. Expect it! Understanding and recognizing fire behavior are critical skills for fire officers to master. As experience levels drop due to the decrease in the number of structure fires, live fire training becomes increasingly important to keep your skills sharp. Remember, your crew is counting on you to make the best call. Their lives are frequently in your hands.

Structural collapse

Another sudden hazardous event for firefighters is structural collapse. Often there is no advance warning prior to collapse. Fire officers have been quoted as saying, "The collapse was a surprise," or, "I never heard any creaking or loud noises until it was coming down all around us." Expect buildings on fire to collapse! Your knowledge of building construction and your study of previous collapse incidents can have a substantial impact on the safety of your firefighters.

Expect buildings on fire to collapse!

Lost or trapped personnel

All too often, personnel become lost or trapped during interior operations. Captain Steve Ruda of the Los Angeles Fire Department stresses that company integrity is the most important factor in reducing firefighter fatalities. Chief Alan Brunacini of the Phoenix Fire Department puts it succinctly: "Firefighters must arrive together, go in together, and come out together." If you look at situations in which firefighters have been lost or trapped, you'll find that in almost all cases company integrity was lost. Captains and lieutenants are the most important link in the chain of accountability. Don't leave your crew, and don't allow any of your crew to leave you!

Company integrity is the most important factor in reducing firefighter fatalities.

Tactical Placement of Engine Companies

We've all heard the old saying, "You can stretch another length of hose, but you can't stretch a ladder." The person most responsible for truck company placement is the officer on the first arriving engine company. It's true that engines must be placed for the best tactical advantage. They need to be close enough to the scene to stretch attack lines, use the deck gun, and have tools readily available. However, 30 feet in either direction won't destroy a good engine spot. The supply line can be stretched, and another section of attack line can be pulled. Ladder companies don't have this luxury. The engine officer's first concern has to be placing the engine in a good position to attack the fire, but don't forget about the first-due truck. Trucks can't support your interior operation if they have to park a block from the fire.

Size of Attack Line

CONSIDERATIONS IN THE SELECTION OF AN ATTACK LINE

What's the best size for an attack line? There are few experts in firefighting who would argue over whether or not laying large-diameter hose (LDH) as a supply line is the best practice. Experience and the science of hydraulics clearly show us that moving larger volumes of water to the pump and reducing friction loss are the options of choice. In contrast, there are a number of theories and significant disagreement on the best size hoseline to pull for an offensive fire attack. Most of these disagreements are based on regional and departmental preferences due to differences in staffing, response times, and the tactical approach to emergencies. The decision on what line to use should address these issues:

- maximum mobility
- speed—how fast and how much

- ease of deployment and pickup
- maximum flow
- personnel available
- water supply available

Practically and theoretically, larger attack hoselines equal:

ADVANTAGES OF LARGER ATTACK HOSELINES

- higher initial attack flow capability
- less friction loss
- longer hose-lay capability
- using more of pump's capacity
- quicker fire knockdown
- less time spent in combat
- more lives saved
- more property saved

THE "PRECONNECT SYNDROME"

Most fire companies struggle with what has been termed the "preconnect syndrome." This syndrome develops as a reflex or an impulsive action based on a continuous habit. The preconnected 1-3/4" attack line is normally the first line off, so it will most likely be called for and pulled even when it is NOT the best line for the fire conditions. Normally, 90% of fires are extinguished with the first two lines pulled, but the engine company officer must consider whether this fire demands a larger line with larger fire flow capacity. Habit should not be allowed to dictate the choice of attack line.

The precon 150 play

The most common preconnected scenario looks something like this:

150 feet of 1-3/4" hose @ 150 psi = 150 gpm

This formula works *most* of the time. It's the easiest and most often practiced pump and line combination. It simplifies pump operation (only one pressure works!), and it eliminates any real thinking during the evolution. It is truly a reflex that we can (and often do) complete in our sleep. It's true that 90% of our fires result in 10% of our losses, but it's also true that 10% of our fires result in 90% of our losses! Engine company officers are responsible for making the call on what attack line selection is the best for this fire, based on an accurate size-up and assessment of resources.

Nozzle Pressure

Nozzle pressure is another component that we have put in the back of our minds, yet it is critical for developing the right response to specific fire threats. The "designed" nozzle pressures are the ones you have memorized for years:

DESIGNED NOZZLE PRESSURES

- Smooth bore nozzle – handline 50 psi
- Smooth bore nozzle – master stream 80 psi
- Combination nozzle (all types) 100 psi

What is the "best" nozzle pressure? Let's consult NFPA Standard 1964, *Spray Nozzles*. The Standard says:

> 2-1.1 Rated pressure for spray nozzles shall be 100 psig. The nozzle discharge rating shall be expressed as a rated discharge at a rated pressure [e.g., 60 gpm at 100 psig].

When we reduce the nozzle pressure, the resulting changes to the fire attack stream are as follows:

RESULTS OF REDUCING NOZZLE PRESSURE

- lower velocity
- less reach
- less impact force
- less heat absorption with fog pattern
- less airflow when used for ventilation
- tendency for hose kinking, causing stream disruption and reduced flow

When we increase the nozzle pressure, we see the opposite effects on the fire attack stream:

RESULTS OF INCREASING NOZZLE PRESSURE

- greater reach
- better penetration
- denser, more opaque fog pattern
- more push to the pattern
- better heat absorption
- more force when the stream hits
- more "action" when it strikes

Dual-Pressure Automatic Nozzles

Most of us have been in a situation in which we are conducting a fire attack in a well-involved, superheated space, and for whatever reason, the hoseline goes limp, or the fire stream is diminished. The screaming starts, and the crews start to back out, heading with a purpose toward the pump operator. There are many reasons for a sudden loss of pressure in a hoseline. Some of these are:

REASONS FOR LOSS OF PRESSURE IN A HOSELINE

- pump transfer valve malfunction
- pump operator error
- incorrect pump operation
- high-elevation losses
- water supply "stolen" by another engine
- long hose lays
- kinked hoseline
- pressure-reducing valves in standpipe systems

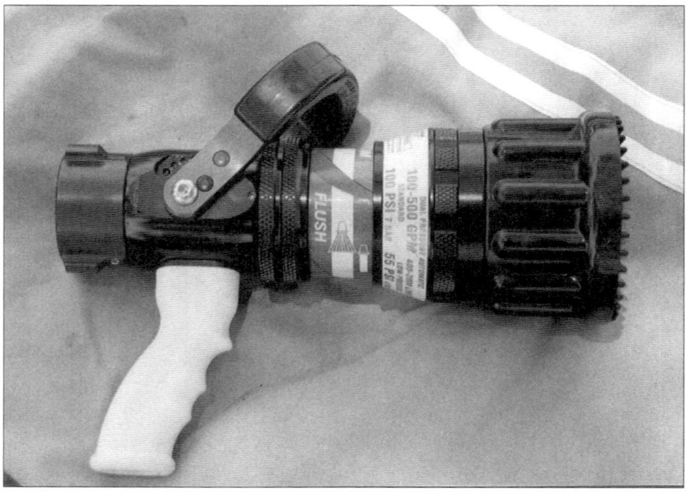

Dual-pressure automatic nozzles such as this nozzle manufactured by Task Force Tips provide for a continuously effective stream during fire attack operations.

With a dual-pressure automatic nozzle, firefighters can quickly switch from standard pressure to low pressure. This allows an efficient and protective stream to continue to be developed by the nozzle. There are many cases in which the use of a dual-pressure automatic nozzle would have made the difference between a safe retreat and a tragic injury or firefighter death.

Truck Company Operations

The Role of Truck Companies

While engine companies play a significant role in the direct attack and mitigation of structure fires, truck companies are the key ingredient on the fireground that dictate engine companies' success.

GOAL OF TRUCK COMPANIES

Truck companies perform a wide range of activities designed to support a single important goal, namely, getting engine company members to the seat of the fire and keeping them there long enough to combat the fire offensively from inside the fire building. Whereas the majority of fires require an aggressive interior attack, there are usually many factors present that will inhibit an engine company's ability to operate safely in untenable heat and smoke conditions. The safety of any occupants in the building is also an issue, since these individuals are not afforded the protection of protective clothing and breathing apparatus.

The role of a truck company is unique, in that all of its members must look at each building involved in fire as a "box" or "cube" that is undergoing a series of rapid dynamics (Figure 4-2). The following factors are all at work:

"BOX" DYNAMICS

- Direct effects of fire consuming the contents.
- Smoke and heated gases pressurizing the atmospheric space.
- Fire extending to and threatening the structural integrity of the framework.
- Fire attack streams placing an unplanned-for weight of pooled water on upper-level floors and on ceilings.

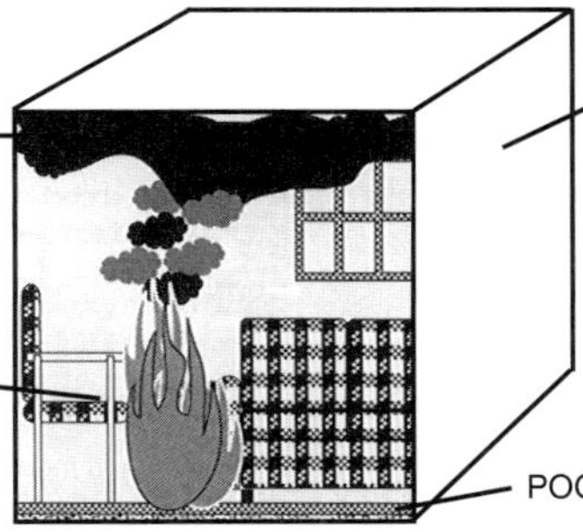

Figure 4-2: "Box" dynamics during a fire.

Because of these "box" dynamics, truck company members must constantly be two steps ahead of the fire. They must always be aware of when and where the fire may be spreading or extending into areas such as the floor or attic space above the fire floor or into concealed spaces in walls and plenums. At the same time, truck company members must be keenly aware of how and where ventilation efforts may be best placed to support engine companies that are attempting to enter the building and occupants who are attempting to exit. As if these matters were not enough, immediate consideration must also be given to the removal of static loads of water forming on the upper floors of buildings or seeping down into unburned contents on the floors below.

One of the most common misunderstandings concerning fire attack operations lies in the myth that to make a fire go out, one must simply keep putting more and more water on the fire via additional attack lines. While this may be true for defensive exterior attack operations, it does not contribute to effective offensive interior attack. Simply put, many things must be done to the "box" physically in order for an incident commander to see a successful knockdown of the fire in a timely manner. The more the "box" is opened up through the efforts of aggressive truck companies, the more the hazard is reduced for those directly involved in interior attack operations.

The more the "box" is opened up through the efforts of aggressive truck companies, the more the hazard is reduced for those directly involved in interior attack operations.

"Truck work" is probably one of the most intelligent forms of manual labor that exists in the workforce today. Well-educated and experienced truck company officers have the ability to provide valuable real-time information to incident commanders concerning the impacts that a fire is having on an involved building and potentially any adjacent buildings. A thorough understanding of the diversity of truck company functions is essential to an officer's success in utilizing assigned personnel to their maximum potential.

Basic Functions of Truck Company Operations

Truck company operations can be categorized into the following 10 functions. Most of these occur sequentially, but they also oftentimes occur simultaneously.

1. Response and placement of truck company apparatus.
2. Forcible entry.
3. Occupant search and rescue.
4. Control of utility services to the building.
5. Placement of ground ladders.
6. Ventilation.

7. Extension check for hidden fires.
8. Salvage of unburned contents.
9. Placement of aerial devices.
10. Provision of elevated fire streams.

Response and Placement

FUNCTIONS OF TRUCK COMPANY APPARATUS

Truck company apparatus perform three basic functions on the fireground. The first and most typical use is in bringing the "toolbox" on wheels to the incident. All forms of truck work require a large variety of tools and equipment that must be easily accessible throughout the duration of the incident. Secondly, trucks provide a means of access to roofs and upper floors in buildings. These areas may be beyond the reach of typical lengths of ground ladders, or the truck's aerial devices may be faster to deploy than ground ladders, which may require the entire truck company to raise. Thirdly, trucks provide a means for delivering fire attack streams, either to upper floors of a building where a viable standpipe system does not exist or during defensive attack operations.

The proper placement of truck company apparatus is critical in getting the best possible use out of the truck on the fireground.

In all of the above cases, the proper placement of truck company apparatus is critical in getting the best possible use out of the truck on the fireground. Fire officers discovered long ago that the nicest thing about engine companies and their attack lines is that they can always be stretched and extended as needed to reach their objective. Unfortunately, this is not true when it comes to truck companies. Every piece of equipment that comes off the truck must be carried, and once an aerial device has reached its specified length, it cannot be extended. Therefore, placement of truck company apparatus must weigh heavily not only on the minds of truck company officers, but also on all officers responding to an incident.

SPOTTING TRUCK COMPANIES

In simple terms, then, the following question needs to be asked when all companies are approaching the fire building: Will aerial devices be needed to support the engine companies at this fire? If the answer is yes, then the address "belongs" to the responding truck companies, and the prime spotting positions that are ideally closest to the building should be reserved for these apparatus—especially if the fully extended length of the aerial device is needed to reach its operational objective. If aerial devices will not be needed, consideration must still be given to allow truck companies to spot in proximity to the building, where they will be unhampered in deploying ground ladders and portable equipment.

Failure to follow these simple guidelines has resulted in many a fire incident in which truck work was extremely delayed due to "truckies" having to carry all of their ladders and equipment from a distance down the street, or aerial devices were painfully missed, all due to the fact that engine companies and other types of apparatus took all of the available positions in the front and/or rear of the fire building. Many departments dictate the response of trucks ahead of engines and other units on their respective alarm assignments, in order to avoid the above scenarios.

Forcible Entry

The very first inhibitor to an aggressive interior attack is the presence of a street-level exterior door that is locked, blocked, or reinforced with security bars. Beyond this challenge lies the fact that one or more locked interior doors may also stand in the way between engine companies and the seat of the fire.

Unfortunately, the various crime rates that exist in any given area will play a large part in the challenges faced by truck companies in overcoming forcible entry problems. The rapid response and arrival of the first-in truck company will enable its members to open up and continue to open up barriers ahead of engine companies that are attempting to advance attack lines into the building.

OPENING UP BARRIERS

In order to effect rapid forcible entry operations, truck company members must be able to size up the degree of the fire problem and decide which level of forcible entry is required. See Figure 4-3 for a comparison of the various types of entry.

Comparison of Types of Entry

Type of Entry	Speed of Entry	Indication	Level of Damage
"Through the lock" entry methods	Slow to moderate	Nothing showing from the exterior or "food on the stove" odor	Low
Conventional entry methods	Moderate to rapid	Working fire with direct access to a door's locking mechanisms	Moderate to high
Power equipment entry methods	Rapid	Working fire with a door's locking mechanisms obscured by bars, scissor gates, roll-up doors, etc.	High

Figure 4-3

In addition to performing forcible entry, truck companies and rescue companies must also consider creating points of egress for engine companies that are operating in the building. In the event of an emergency, these additional doors and/or windows can serve as a lifesaving means of escape for firefighters or a quick means of access for rapid intervention teams to effect the rescue of downed or trapped firefighters.

CREATING POINTS OF EGRESS/ACCESS

Occupant Search and Rescue

Once initial forcible entry is accomplished, one or more truck company members are typically tasked with effecting a primary search of the fire building behind the advancement of the initial attack line(s).

CONDUCTING A PRIMARY SEARCH

The initial or "primary" search consists of a rapid check of occupied floor areas for occupants who may have attempted to exit the building before becoming disoriented, overcome by smoke and toxic fire gases, or cut off by the fire itself. Of greater difficulty to locate are those occupants who were overcome prior to being aware of the fire or those who choose to hide in closets, bathtubs, etc., out of fear of their present circumstances. Due to the fact that most primary search operations begin in the early phases of an initial attack, heat and smoke conditions typically are at their worst levels. This is usually because other truck company members are in the process of preparing to ventilate the building, which oftentimes requires making access to the roof and taking equipment aloft to open the roof. Needless to say, most primary searches, then, are conducted at near-zero visibility, leaving those searching to rely heavily on their tactile and hearing senses to locate potential victims.

Once victims are located, these same members must quickly remove these occupants out of the building in order to improve their chances for survival. In larger buildings, such as multistory or high-rise apartment buildings, occupants may simply be relocated to other floors of the building (above or preferably below the fire floor) that are void of toxic fire gases and smoke.

As discussed in Module 2, completion of the primary search is an important tactical benchmark. When an "all clear" is given following the completion of the primary search, this along with the completion of several other key objectives leads the incident commander to the conclusion of the initial action plan and allows the incident to be declared "under control."

CONDUCTING A SECONDARY SEARCH

In addition to the primary search, truck company members are also typically tasked with conducting a secondary search of the building as well. The methods associated with this search differ greatly from those involved in the timing and execution of a primary search. A primary search is conducted while fire attack and truck company support functions are occurring on a working fire. In contrast, a secondary search is conducted only after three fireground events have occurred:

- All active fire has been knocked down.
- The building has been ventilated.
- Adequate lighting (as needed) has been set up inside the building.

Since a primary search is conducted in near-zero visibility conditions, there is a possibility that some occupants will not be located. Therefore, a secondary search conducted under the conditions above allows a much more thorough examination of all potentially occupied areas. This task is more manageable once visibility issues associated with interior smoke conditions are improved through the aggressive efforts of the remainder of the truck company performing ventilation functions. It is important to note that when the second "all clear" benchmark is reached following the secondary search, the building is formally declared to be void of any remaining occupants. Consequently, the task of conducting a secondary search brings with it a great deal of responsibility. There is nothing worse than locating a victim(s) later during salvage and overhaul operations.

Utility Control

The majority of truck company work is "invasive" in nature. As discussed earlier, the basic role of the truck company is to "open up" a number of structural areas of a fire building in order to enhance the attack efforts of engine companies. Typically these structural areas include roofs, ceilings, walls, and floor spaces. As in most buildings, these same areas contain a significant array of wiring and plumbing networks that provide electrical service and heating fuels throughout the building. For obvious safety reasons, it is imperative that rapid and accurate control of utilities take place. Truck company members must be well versed in identifying the types of utilities that are servicing a fire building. Coupled with this task is the ability to locate rapidly where the exterior or interior shut-off locations are and the best way to terminate utility services at the place where they enter the building.

LOCATING UTILITY SHUT-OFFS

For most occupancies that involve single- or multifamily dwellings (excluding high-rise occupancies), locating and shutting off utilities is a relatively simple task. For large commercial and industrial buildings, utility shut-offs may be located inside the building in areas such as electrical control rooms or vaults. Effecting shut-off of these utilities, particularly electrical power, may also pose some safety considerations due to a large amount of voltage supplying the building's energy needs.

Once a fire has been knocked down, consideration must be given to controlling water services to the building as well. Oftentimes water pipes within the building may break due to fire conditions or attack operations, resulting in dangerous static weight loads on upper floors and ceilings.

Placement of Ground Ladders

A well-equipped truck company should carry a wide variety of ground based ladders that will allow the company to access all areas of a fire building.

Examples of such ladders include:

TYPES OF LADDERS CARRIED BY A TRUCK COMPANY

- 35-foot extension ladders for access to roofs and upper-floor windows.
- 24-foot extension ladders for access to roofs and upper-floor windows.
- 20-foot straight and roof ladders for access to roofs and upper-floor windows and to provide a stable climbing surface when cutting ventilation holes on steep-pitched roofs.
- 16-foot straight and roof ladders—used for the same purposes as listed above for 20-foot ladders.
- 12-foot roof ladders—also used for providing a stable climbing surface when cutting ventilation holes on shorter, steep-pitched roofs.
- 14-foot extension ladders for access to interior attic space and roof scuttle hatches.
- 12-foot combination step and extension ladders for special access situations across fence lines, etc., and when a free-standing ladder is needed (i.e., salvage operations, water chutes, etc.).

IMPORTANCE OF PLACING LADDERS EARLY IN THE INCIDENT

A majority of truck company functions, especially ventilation operations, require the fire building to be adequately "dressed" with the appropriate number and type of ground ladders. Once these ladders are in place via initial and subsequent arriving truck companies, they will provide ample means of staying ahead of the fire, while also providing above-ground points of egress for engine companies and occupants inside the building and for truck companies operating on the roof.

The key ingredient for successful overall fireground operations lies in the ability of truck company officers to ensure that a sufficient number of ladders are placed early in the incident, especially when truck companies initially arrive and are intact to throw ladders that require two or more members.

Ventilation

Ventilation operations lie at the core of a truck company's purpose and existence. The reason for this is that properly timed and executed ventilation operations will have the greatest impact on fire attack efforts that are simultaneously being carried out on the inside of the building. Even more so, ventilation operations directly affect the degree of safety that exists for engine company members and building occupants, with regard to the mitigation of potential flashover and backdraft conditions as described in Module 1.

TYPES OF VENTILATION OPERATIONS

Ventilation operations can be categorized into several simple terms:

- offensive "heat hole" operations
- defensive "strip" operations
- positive pressure operations

Offensive "heat hole" operations

A "heat hole" in the roof will dramatically affect the vertical ventilation of heated gases and smoke.

The "heat hole" operation is probably the most common "bread and butter" operation for a truck company, since most initial attack operations involve a working contents fire requiring an aggressive interior attack. If smoke and heat conditions involve the uppermost floor of a building or a one- or two-story single-family dwelling, the truck company will mirror the offensive attack by opening the roof and ceiling as close to the seat of the fire as possible (Figure 4-4). This "heat hole" in the roof will dramatically affect the vertical ventilation of heated gases and smoke by allowing these products of combustion to exit the building while following their normal path of convected heat transfer. As a result, interior conditions for engine companies and building occupants will drastically improve in terms of increased visibility and decreased heat levels.

Figure 4-4: Use of a heat hole during interior fire attack operations.

Defensive "strip" operations

COMBATING FIRES THAT HAVE EXTENDED INTO ATTIC SPACES

Unfortunately, many contents fires do not remain confined to the occupied spaces in the building. Subsequently, single-story or upper-floor fires may extend into a common attic space and spread laterally in all directions. When this occurs, the structural integrity of the roof system itself becomes threatened, resulting in eventual failure and collapse, depending on the materials and methods used to construct the roof. When roof collapse occurs, engine companies and occupants face the reality of being trapped under such a collapse and potentially succumbing to the effects of trauma or suffocation.

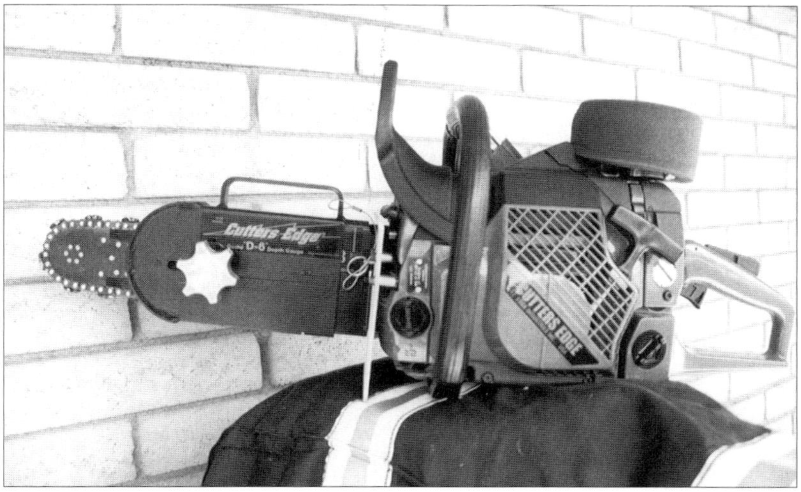

Chainsaws are one of the most effective pieces of equipment for cutting heat holes and strips in roofs.

Truck company officers must quickly recognize when fires have extended to these attic spaces and must in turn cut off the lateral spread of the fire, while coordinating with engine companies on the interior to knock down the fire from below. Defensive "strip" operations simply involve cutting a long, narrow, rectangular hole in the roof ahead of the fire, in an effort to introduce a fresh oxygen source for the fire when it reaches the strip. It is here that three events occur:

HOW A DEFENSIVE "STRIP" WORKS

- The fire exits the length of the strip instead of continuing to spread laterally.
- Truck company members pull ceiling below where the strip is located on the roof.
- Engine companies knock down the existing fire in the attic.

Positive pressure operations

When working fires occur on lower-level floors, the opportunity for vertical ventilation does not always exist. As a result, dangerous levels of heat, toxic gases, and smoke accumulate and are unable to follow a vertical path of convection out of the building. When this occurs, horizontal ventilation is required. This type of ventilation is greatly enhanced by introducing a controlled airflow into the building and providing a means for the products of combustion to be exhausted out of the building opposite the side where the airflow is introduced.

HOW POSITIVE PRESSURE VENTILATION WORKS

Positive pressure ventilation (PPV) is accomplished by placing power equipment fans in specific configurations at a designated opening (typically a doorway). As the flow of air is introduced from the fans, atmospheric pressure

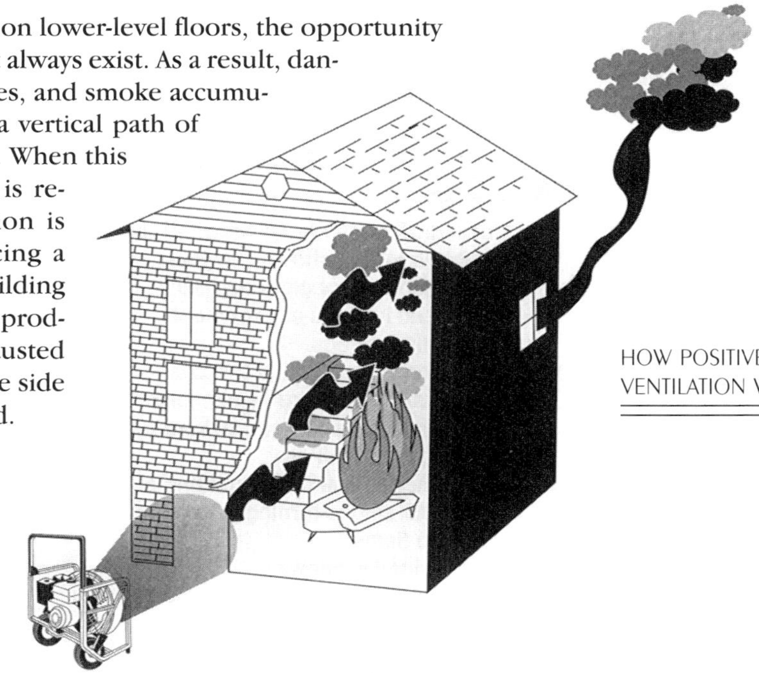

Figure 4-5: Use of positive pressure ventilation.

Gasoline powered blowers provide for maximum portability and ease of placement for positive pressure operations.

begins to build inside the structure. When an exhaust opening is created (via windows or another door), then the pressure acts to displace the products of combustion in a fairly rapid manner. PPV is also a useful tool in conjunction with vertical ventilation. When used in this manner, the heat hole acts as the exhaust opening to remove the residual heat, smoke, and carbon monoxide levels that are typically present during overhaul operations. Ventilation techniques will be covered more thoroughly in Module 5.

Extension Check for Hidden Fires

The key to successful identification of hidden fires lies in the truck company officer's ability to understand the construction of the fire building.

As discussed earlier, "contents fires" can typically progress to "structural fires" when the fire extends into a variety of hidden spaces. These spaces exist above and below floor levels, in attic spaces, in fascia areas, behind walls, and anywhere that utility services, chase ways, or ventilation pipes transcend the structure. The key to successful identification of hidden fires lies in the truck company officer's ability to understand the construction of the fire building. It is from this understanding that buildings are opened up from the inside out, in order to allow engine companies to knock down fires that can spread rapidly in these concealed spaces.

It is important to remember from a safety standpoint that any fire left unchecked in a concealed space lends itself to threatening the structural integrity of the building. When engine companies are required to operate on or underneath floors or in attic areas, where fire has captured void spaces, such conditions can place these companies in danger of facing a structural collapse. A truck company officer always thinks ahead of the current location of a contents fire and trains company members to open up all adjacent void spaces, preferably while engine companies are getting a knockdown on the contents fire itself.

Salvage of Unburned Contents

Salvage operations involve several areas. In order of priority, they include:

SCOPE OF SALVAGE OPERATIONS

- Assessment of static water loads on upper floors and ceilings and the overall stability of the fire-weakened structure.
- Removal of static water loads from "nonporous" floors (i.e., concrete floors and tile or linoleum covered wood floors) and ceilings (i.e., lath and plaster ceilings).
- Protection of unburned contents from subsequent water damage from attack lines or sprinkler heads.

Due to the fact that aggressive interior fire attack operations require engine companies to provide a gallon-per-minute flow of water that exceeds the amount of heat being released by the fire, a considerable quantity of water may end up inside the building after knockdown of the fire has been accomplished. Additional quantities of water may be introduced into concealed attic and floor spaces to extinguish hidden fires.

As a result, truck companies often find themselves as the key players in finalizing the loss of property that occurs due to fire department actions. In conjunction with this, truck companies are also tasked with making the building safe for overhaul and fire investigation activities to occur, by removing static loads of water that the building was not designed to carry.

Truck company officers must be able to direct salvage efforts quickly according to specific priorities, in order to save as much property as possible.

While little can be done to salvage contents that have already been damaged by fire, much can be done to prevent unburned contents from being damaged by water. Truck company officers must be able to direct salvage efforts quickly according to specific priorities, in order to save as much property as possible. By following the priorities listed below, many departments have utilized successful salvage operations as a very positive public relations tool with owners and occupants of dwellings and businesses.

SALVAGE PRIORITIES

Salvage priorities in order of "save first" sequence

- Habitational
  - home electronics
  - furniture and antiques
  - artwork
  - clothing
- Office buildings
  - electronic office equipment
  - unprotected files
- Commercial/industrial
  - accounts receivable
  - machinery
  - finished product
  - raw materials

Placement of Aerial Devices

Aerial ladders can be used to support fire attack operations in the following ways:

USES FOR AERIAL LADDERS

- Provide access to roofs beyond the reach capability of ground based ladders.
- Effect rescue from upper-floor windows.
- Provide points of upper-floor emergency egress for engine companies.

CONSIDERATIONS FOR USING AERIAL DEVICES

The decision to use an aerial device for any of the above applications is an obvious one when vertical height requirements exceed the reach capability or availability of ground based ladders. A prudent truck company officer will commit all ground ladders to objectives within their reach, while reserving the aerial device for lengthier objectives. Another consideration includes the staffing levels that are on the responding truck companies in relation to the number of ground ladders that need to be thrown. One of the advantages of using an aerial device is that it essentially requires only one truck company member to place the ladder, while freeing up other members to throw additional ground ladders or accomplish other fireground truck company duties.

Provision of Elevated Fire Streams

The final service provided by truck companies involves the capabilities for directly delivering fire attack streams for offensive or defensive operations.

OFFENSIVE OPERATIONS

For offensive operations, the aerial device on the truck company apparatus can be utilized as an exterior "standpipe system." In this manner, attack lines can be extended into a number of floors on a multistory building through the use of waterway features built into the aerial device. Another option is to lay a larger-diameter line up the center of an aerial ladder and utilize gated wye-type appliances to supply attack lines. These lines then can be advanced through adjoining windows or fire escape access points in the building. Use of an aerial device in this manner is especially helpful to engine companies for buildings that lack a standpipe system or whose system is out of service or unreliable.

DEFENSIVE OPERATIONS

For defensive operations, aerial devices provide an efficient means of delivering heavy streams from elevated positions adjacent to or above the building. From a tactical standpoint, this is one of the most advantageous means of delivering the gallon-per-minute (gpm) flow required when attacking building fires that are fully involved. Furthermore, once fires have collapsed the roof system of a structure, elevated heavy streams have the ability to knock down the fire quickly, depending on the gpm flow in relation to the contents and structural load that is burning.

DEALING WITH BURNING EMBERS

An inherent problem associated with defensive attack operations is the burning embers that emit from the building, which can become airborne and travel great distances. These embers can subsequently ignite fires on adjacent roofs, start fires in nearby vehicles with open windows, and start grass fires that are often miles away. One or more elevated fog streams can be positioned on the downwind side of the building and utilized to extinguish these burning embers before they can reach combustible targets. When exposed property ignites in close proximity to aerial devices, aerials can often be rotated to direct the elevated stream onto the exposed property and quickly extinguish the fire.

Seat Assignments for Truck Company Responsibilities

Seat assignments for truck company responsibilities dramatically increase the overall efficiency of the company. Unlike engine companies that tend to function as a group on the fireground, truck companies tend to fragment their operations, in order to provide a greater number of simultaneous support functions.

VALUE OF PREDESIGNATED SEAT ASSIGNMENTS

Truck company officers who predesignate seat assignments at the beginning of every shift will likely see their company operating at a high degree of productivity both collectively and independently. In order to provide the best utilization of personnel, officers must identify the staffing level of their company in relation to the truck company functions listed earlier. As such, these functions can be divided into two initial attack operational categories:

Inside Operations

- forcible entry
- occupant search and rescue
- extension check

Outside Operations

- utility control
- placement of ladders
- ventilation

Based on these categories, seat assignments can then be made in accordance with the number of members assigned to the company. Typical examples are listed below:

6-Member Staffing

- 2 members for inside operations.
- 3 members for outside operations.
- 1 Officer assigned as the Roof Division Supervisor.

In the above example, tool assignments would include:

- 1 "irons" member to effect forcible entry and search.
- 1 "hook" member to effect extension checks and assist with the search.
- 1 "cutter" to throw ladders and cut the heat hole or strip.
- 1 "puller" to throw ladders and open the heat hole or strip.
- 1 "utilities" member to throw ladders and control utilities. Note: Once these operations are complete, this member could also initiate interior salvage operations early in the incident—a distinct advantage to minimizing property loss due to fire department operations.

5-Member Staffing

- 2 members for inside operations.
- 2 members for outside operations.
- 1 Officer assigned as the Roof Division Supervisor.

In this example, tool assignments would include:

- 1 "irons" member to effect forcible entry and search.
- 1 "hook/utilities" member to effect utility control and extension check and assist with the search.
- 1 "cutter" to throw ladders and cut the heat hole or strip.
- 1 "puller" to throw ladders and open the heat hole or strip.

4-Member Staffing

- 2 members for inside operations.
- 2 members for outside operations.

Here, tool assignments would include:

- 1 "irons" member to effect forcible entry and search.
- 1 "utilities/hook" member to effect utility control and extension check and assist with the search.
- 1 "cutter" to throw ladders and cut the heat hole or strip.
- 1 "puller" (the Company Officer) to throw ladders and open the heat hole, in addition to serving as the Roof Division Supervisor.

In all of the examples listed above, it is important to note that the entire truck company should remain intact upon arrival at the incident, if staffing levels are such that all members will be required to throw the ladder(s) necessary for vertical ventilation work to take place.

Also, in all cases the Truck Company Officer may retain all of his or her members in order to conduct more complex ventilation operations on the roofs of larger commercial and industrial buildings. In this case, coordination with a subsequent arriving truck company or engine company would be required in order for inside operations to be accomplished.

Once all initial inside and outside operations are completed, the company should at that point be prepared to conduct subsequent ventilation operations using positive pressure ventilation methods and also initiate salvage operations on the inside of the building.

Summary

Engine company officers are the officers that are most responsible for extinguishing the fire. The support activities are important, but all fireground tasks are centered around improving the ability of the engine crews to locate, confine, and extinguish the fire. The engine company officer has a heavy responsibility for the safety of the interior crews. Flashover, collapse, and firefighters becoming lost and trapped are common occurrences that often result in death. Understanding fire behavior, being aware of predominant building construction types, and demanding company integrity can increase the level of safety and survivability for engine company personnel.

The success of engine company operations is directly attributable to the rapid and timely execution of aggressive fireground truck work. A considerable number of support functions must occur systematically and often simultaneously if structural fire problems are going to be mitigated both rapidly and safely. Using predesignated seat assignments for truck company members can increase the efficiency of the company as they execute their responsibilities on the fireground.

MODULE

5

Standards of Fire Attack: Offensive Operations

Objectives

After completing the reading and activities in this chapter, students should be able to:

OBJECTIVES

- Identify and understand the role and responsibilities of the company officer when assigned as a division or group supervisor on all-risk incidents.
- Implement tactical operations related to interior fire attack objectives.
- Implement tactical operations related to occupant search and rescue objectives.
- Implement tactical operations related to vertical and positive pressure ventilation objectives.
- Implement tactical operations related to property salvage objectives.

Offensive Fire Attack

WHEN OFFENSIVE FIRE ATTACK IS INDICATED

Offensive fire attack is the most common fireground activity. This strategy is indicated when the resources arriving on the scene have the capability to perform life safety and rescue operations, initiate fire/incident stabilization, and take action to conserve property. Offensive fire attack is the strategy whereby search operations are performed. Rescue operations are indicated when the search function is successful. A second focus of offensive fire attack is ventilation. Ventilation can improve visibility and reduce the heat for firefighters and any trapped occupants. Coordinated ventilation techniques can also assist in confining the fire and reducing property loss. Most salvage operations will be performed in the offensive mode as well.

The coordination of interior fire crews, ventilation, search, exposures, rapid intervention, and salvage activities requires a focused and confident incident commander directing equally competent fire companies in order to be successful. This module will discuss the coordination and activities found on offensive fire attack incidents.

Offensive Considerations

Many times, the best method to stabilize and control an incident is a rapid, interior fire attack. At other times, the best approach is to take the heat out of the main fire with a heavy stream before starting the interior fire attack. There are a number of combinations of these tactics. The decision as to which approach to use should be based on the occupancy, use, construction, egress, and level of fire activity.

Inside/Out Attack

ADVANTAGES AND DISADVANTAGES

Offensive operations can be divided into two basic attacks: "inside/out" and "outside/in." In an inside/out attack, the attack crew members enter the structure, extend hoselines into the fire area, get behind the fire, and push the fire out a predetermined opening. The exit hole may have to be opened by another crew member or company and coordinated on the orders of the interior crew. In the inside/out attack, great care is taken not to push fire through the structure but to direct it out a door, window, or other opening. The advantage of this attack is that it limits fire extension. The disadvantages are that interior crews take a beating from the heat, fire attack time is delayed until the nozzle is behind the fire, and flashover may occur due to the delay.

Outside/In Attack

In an outside/in attack, the fire attack team initiates a quick knockdown from outside the building or from the entryway. Sometimes this is best done with a deck gun or heavy line. The advantage of this approach is that it prevents flashover by knocking down the fire. The disadvantages are that heat and smoke may be pushed toward occupants, the fire may extend into other uninvolved rooms or spaces, and there will be more water damage due to the fire's being confined within the structure and not pushed outside.

Apparatus Placement

Apparatus placement at offensive incidents is of critical importance. Deciding who "owns" the address is up to the first-in officer. In situations in which an aerial device will be used for rescue or vertical ventilation, provisions must be made for giving the truck the front of the occupancy. This placement never happens by chance. It requires active consideration by the first-in officer. Conversely, if the situation requires a master stream or heavy hoselines immediately, and the use of the aerial device is probably not necessary in the front or center of the building, then the engines get the front, and the truck needs to find a corner spot to cover exposures or to provide a later elevated stream operation. Having subsequently arriving engines stage in a flexible location, near a water source, will allow the incident commander to bring them in to the best tactical position when the need arises. Too often, company officers overcommit their staging location too close to the initial arriving companies, and the IC is forced to bring apparatus in and utilize their crews in a less-than-perfect spot, because there is no other option without moving companies or relocating hoselines.

Primary and Secondary Searches

Both a primary and a secondary search are conducted during offensive operations. A primary search is the initial search made by an assigned company to find live victims. This type of search concludes with the tactical benchmark, "Primary search complete." The secondary search is a more detailed, methodical search, in which deceased victims are sought. If no victims are found, the search transmission terminates with the benchmark, "All clear." If fire conditions prevent a thorough primary or secondary search, then the message, "No all clear," is transmitted.

Ventilation

Offensive operations are also characterized by the use of exterior vertical and interior positive and negative pressure ventilation techniques. These techniques will be discussed in detail later in this module. In a defensive fire, there are no ventilation operations. The exception is the use of a defensive trench or strip ventilation operation to stop the fire from extending into an exposure.

Incident Command Organization

An offensive attack is much more rapid and aggressive than a defensive attack. Time is as much the enemy as the fire. Successful company officers are those who can quickly organize their assigned personnel and rapidly execute the required tasks. The incident command organization should be developed to supply the necessary resources to provide fire attack, search, rescue, ventilation, exposure control, and salvage operations. In addition, support functions must include command, sector/division/group officers, water supply, safety, and rapid intervention.

Managing Searches

A critical element of an offensive fire attack is managing and conducting search operations. Search operations are conducted on occupancies when there is a probability that live victims exist. Defensive fire conditions will typically not warrant an interior search for live victims.

Search Strategies

As noted above, there are two basic search strategies: primary and secondary. The difference between the two is the condition of the possible victims that you are trying to find.

PRIMARY SEARCH

A primary search is a rapid, systematic search to locate savable victims. There is always a logical place to start and a logical place to stop a primary search. There is also a time frame beyond which the search should not or need not be conducted any longer.

SECONDARY SEARCH

The secondary search is a slow, methodical search to look for and locate dead victims. It should be noted that not every reported "person trapped" will automatically result in a search team's locating a live victim. Most of the time, the person has exited the structure, was not in the involved structure in the first place, or succumbed to heat and smoke before the search was ordered. It should be noted that the secondary search is not merely a more thorough search of the same locations previously covered by the primary search team. Unless the primary search process is flawed, the chance of finding victims the second time around who were missed the first time does not increase.

Types of Searches

There are three main types of searches:

- standard search
- team search
- oriented search

The advantages and disadvantages of each are outlined below. A detailed description of the third method, the oriented search, is given later in this module.

Standard search

The standard search technique can be conducted in several variations. The first method is that one member goes left, another goes right, and they meet somewhere near the middle of the room. A variation is that both members form a chain and follow the room from wall to wall.

ADVANTAGES AND DISADVANTAGES OF THE STANDARD SEARCH

There are some advantages to the standard search. First, it conforms to OSHA's two-in/two-out standard. Secondly, all members are responsible for some search area.

The main disadvantage to this type of search is that the officer involved loses focus while searching under beds and sweeping the room. The safety of the crew is no longer the foremost objective. Questions such as, "What is the fire doing?" and, "Can every person with a radio be immediately located within the structure?" are temporarily set aside. Instead of thinking about the safety of the crew, the location of the fire, and the coordination of the overall operation, the officer is focused solely on searching. This can put search teams in danger. The other major disadvantage to the standard search is that only one area can be searched at a time.

Team search

An alternative to the standard search technique is the team search. This method utilizes a search rope, hoseline, or other fixed object. The rope or hoseline is connected to an anchor point in a safe area. Members work off this guide and search using some predetermined pattern.

ADVANTAGES AND DISADVANTAGES OF THE TEAM SEARCH

One advantage of the team search is that crews remain together. Another advantage is that with an orientation to the rope or hoseline, all members can participate in search efforts. The team search can cover a considerable amount of area very rapidly.

A disadvantage of the team search method is having a sole reliance on the rope or hoseline. In a large area search, crews can get in too deep, and their only lifeline is the rope or hoseline. In actual tests and demonstrations, search crews spend about as much time manipulating the hoseline or rope as they do searching.

Oriented search

The oriented search is defined as a method of search that utilizes one "oriented" person (team leader) and one or more searchers. It is a method of search that allows the focus to be split into two directions:

- the safety of the search team
- the actual search

ADVANTAGES AND DISADVANTAGES OF THE ORIENTED SEARCH

The primary advantage of the oriented search is that the safety of the search team is maintained at all times. Additionally, the searchers are allowed to focus exclusively on finding victims. They are not burdened with hoselines or ropes, so searches are conducted faster. This equates to more lives saved, and the exposure of the search team to the hazards is reduced.

One disadvantage of the oriented search method is that this method requires a lot of concentration on the part of the oriented person or team leader. This individual has to be a good thinker and has to think way ahead of the current situation. The team leader also must be aware of the location and position of the fire.

In the 1980s, tests were conducted comparing the standard search with the oriented method. The tests revealed that firefighters conducting an oriented search were much more confident and focused on the search, because they knew that someone else was looking out for their safety. Oriented searches were thus conducted faster than standard searches.

Oriented searches are conducted faster than standard searches.

Responsibilities of Command

PRIORITIZING SEARCH OPERATIONS

The incident commander should prioritize when search operations are to begin. This decision should be based on a comprehensive size-up and evaluation of fire conditions and available resources. If resources or available staffing is low and the choice is between putting out the fire and searching, PUT THE FIRE OUT! With adequate staffing and equipment, initiate search operations.

In developing an offensive attack plan with limited resources, the logical sequence of activities assigned at a fire or tactical progression is as follows:

- attack
- ventilation
- search
- backup
- salvage/overhaul

If staffing is not an issue, consider this progression:

- ventilation
- attack
- search
- backup
- salvage/overhaul

Mission of Search

COORDINATING SEARCH EFFORTS

The mission of the Search Sector/Division/Group is to *coordinate* search efforts in the area assigned. The coordination involves assigning, tracking, and supporting the teams deployed to search. The officer assigned to the search operation has three responsibilities:

- crew safety
- coordination of search efforts
- keeping Command informed

ASSUMPTIONS MADE BY INCIDENT COMMANDER

Keeping the incident commander informed of the position, progress, and needs of the Search Sector is vital. Apart from any communications to the contrary, the incident commander must make some assumptions concerning this sector. The first is that Search is actually coordinating the efforts of multiple crews. The second assumption is that Search has developed a plan. The plan must include where to search first and where to search last. Third, Command must be reasonably confident that Search will focus on safety first. Finally, Command should assume that when Search changes floors, Command will be notified.

ASSUMPTIONS MADE BY SEARCH OFFICER

Search must also make some assumptions. First, Search should believe that if Command receives some information concerning the search, then Command will relay that information to Search. Search should also be confident that Command has assigned a backup and rapid intervention team and that Command is watching for changing conditions.

Other responsibilities of the search officer are to maintain focus on the mission and to request adequate resources to complete the job. The search officer's hands should not be sweeping under beds or in bathtubs. The search officer must maintain a broad enough vision to focus on the search and a narrow enough vision not to run the incident.

If Search believes that there are still savable victims in the building/area and that the crew(s) cannot cover the area, then Search should ask for more help and should again coordinate tasks.

Where to Search

Normally, the search should start as close to the fire as possible.

The first rule in search is to determine where to search first and where to search last. Normally, the search should start as close to the fire as possible, where savable victims could be, and then should work out and up from there. The exception to this rule is a fire on the first floor, when the time indicates that savable victims are above the fire. In all cases, the first-arriving officer assigned to Search must read the building for search. This requires identifying the most appropriate way into the structure and visualizing the probable locations where savable victims are most likely to be.

Under most circumstances, Search should follow the hoseline into the building. This method usually works, as fire attack is normally assigned first. The other benefit of following the hoseline into the structure is that it assures that a line is stretched between the fire and the search team. It also reduces the potential for the search team to get hit with attack streams.

After gaining access into the structure, the search officer must initiate the plan. On a working fire, conditions will dictate that the search start as close to the fire as possible. This is often the least tenable or survivable space for a trapped victim. The crew should follow the line in up to the attack crew and then work back toward the point of entrance.

In a multistory occupancy or at a fire where those above the fire are the most threatened, the search crew should locate the stairway, if possible from outside. They should then follow the line in and ascend the stairs.

Limiting the Primary Search

We have been taught that a search concludes when all victims are located. There has never been a definable limit on the length of a search. The obvious flaw in this logic is that the fire conditions and structural disintegration are ongoing, continuous processes during the search. There has to be a more quantifiable limit for fire commanders to reference when deciding to terminate an unsuccessful search process. Chief Skip Coleman of the Toledo, Ohio Fire Department recommends the maximum time frame that a primary search should last is 15 minutes. This limit is developed using human physical factors, moral factors, and common sense.

The maximum time frame that a primary search should last is 15 minutes.

Under fire conditions that are defined as immediately dangerous to life or health (IDLH), it is reasonable to assume that fire companies should conduct a primary search for approximately 15 minutes based on the following considerations:

- The duration of one SCBA bottle is 15 minutes. After this time, crews are in danger of becoming overextended with no backup air supply.
- The human body can only survive for 4-6 minutes at 15% oxygen.
- The body can sustain temperatures of 180°F (82°C) for no more than 4-6 minutes.

Based on these facts, it is reasonable to limit searches to 15 minutes. For victims, this is a double or nearly triple buffer factor provided under IDLH conditions, while being careful not to jeopardize crew safety.

In better than IDLH conditions, searches can be conducted when staffing allows or when other emergency responders are available. The maximum time frame can be lengthened when the survivability is increased. The search officer should plan the evolution based on the conditions and the maximum time frame in mind.

Maintaining the Focus of the Search

There are many activities that have traditionally been associated with the search, which actually detract from the search team's ability to maintain their focus on just one thing—locating victims who are unable to exit the hazardous area. When the search becomes a rescue, the search process is compromised. When the search team is stretching a hoseline, the effort expended also detracts from the search.

When the search becomes a rescue, the search process is compromised.

Search vs. Rescue

"Search and rescue" has been referred to as one task. They are and should be two separate and distinct operations. Search is locating those trapped or unable to exit. Rescue is removing those who have been located out of the hazard zone. All sector assignments are built on focus. When a search team locates a victim, what happens to the focus? It immediately changes from finding the victim to bringing the person out of the building. While the victim is being removed from the structure, the search for other victims is no longer a focus of the operating crews. Search comes to a screeching halt until the crews transfer the victim to EMS and reenter the hazard zone for another look.

DUTIES OF A RESCUE GROUP

Obviously, if there is only one reported victim missing and all other occupants are accounted for, then Search can and should remove the located victim. On an incident with multiple victims or an unknown number, a large area to search, and the probability of multiple removals, a Rescue Group should be established. Every time a person is located, the Rescue Group should be assigned and deployed to remove the victim to outside the hazardous area. Some of the occupancies where this situation may be confronted include:

- hotels and motels
- high-rise buildings (residential and commercial)
- occupied stores (strip malls and enclosed malls)
- night clubs and restaurants

In the above occupancies, Search should limit their activities to locating victims. As victims are found, searchers should bring them to a common area, such as a hallway, entrance, foyer, or stairwell. The search team should then request an appropriate number of rescuers. The rescuers should receive the victim from the search team and then remove the person to the EMS triage or treatment area. The searchers should immediately go back and complete the search objectives.

Hoselines for Search Teams

Search teams should use a hoseline only under specific conditions and only with additional support.

It has been taught that all search teams must have a charged hoseline in order to conduct a search safely. In this course, we advocate using a hoseline only under specific conditions and only with additional support. The first question that a company officer should ask when assigned to search operations in a residential occupancy is, "What are the fire conditions?" If there are active fire conditions contributing to an IDLH environment, then the potential for live victims in the space is adversely affected.

After the fire conditions have been assessed, capability or resources should then be considered:

- How many members are in the crew?
- How many members does it take to stretch and extend the proper hoseline in the occupancy?
- What physical energy and air supply are expended just getting the line in place before the search activities even begin?

Finally, the focus of the search team should be evaluated:

- How many members are left to focus solely on locating victims?
- Where is the focus?
- Are the crew members giving the victim(s) their best effort?
- Are they searching as you would like crews to search for your family members?

If there is an active fire attack team with a backup line in place, the search team is free to enter the structure and conduct an unencumbered search.

COMMERCIAL OCCUPANCIES

An exception to the "hoseline" rule is in commercial occupancies. The backup team can pull the line, and then the search team can work off that line. When a Search Sector/Group is assigned, a Backup Sector/Group should also be assigned. The Backup Group's task is to shadow Search and assure that they are safe and that egress is continuously available.

Backup Line

WHERE TO POSITION THE BACKUP LINE

The placement of a backup line is critical in providing search and fire attack personnel the proper margin of safety. On a first floor or upper floor, the backup line protects the path of egress for the search crew and provides protection between the fire and the occupants. The backup line for a basement fire should be positioned at the top of the basement stairs with additional slack line pulled. For an attic fire, the backup line should be positioned at the bottom of the attic stairs with extra slack line pulled.

Searching With a Tool

Truckies and rescue companies have always been taught to carry a short tool, such as a halligan, an O-tool, or a short hook, to use as a sweeping extension and to force egress in case of entrapment. However, interviews with hundreds of firefighters, coupled with actual demonstrations and timed search evolutions with and without tools, have shown that the teams with the tools conduct much slower searches. They also spend a large amount of time and energy carrying, dragging, dropping, picking up, and changing hands for little or no benefit. Among firefighters who state that they always carry a tool in case they need to force a means of egress, very few have ever had to use the tool for that purpose. In light of these statistics, the question must be asked, "Is the cost worth the benefit?" Only in the oriented search is there a definite benefit. Here, the oriented person can and should have a tool. In this situation, the search is not slowed down, since the oriented person is not actually sweeping and searching.

Teams with tools conduct much slower searches.

Searching With a Thermal Imager

Thermal imaging cameras, or TICs, are an incredible tool for search teams. In departments fortunate enough to have multiple TICs, several members should be provided with an imager. The neckstrap option on a lightweight model seems to work best. The larger, handheld devices work very well for the company officer or oriented person to carry. Using a TIC for "active firefighter accountability" inside the hazard zone is one of the greatest improvements in firefighter safety in several decades.

Searching for Life or Fire

SENDING CREWS ABOVE THE FIRE

Some larger departments send crews above the fire to search for both life and fire. The second-due ladder company is frequently assigned this function. This practice can lead to a rapid finding of trapped victims and can give the incident commander early indication of fire extension and vertical travel. When encountering a victim above the fire, the crew should treat the situation as any other search. They should call for a rescue team to bring out the victim, unless it is confirmed that no other victims are missing. If the team finds fire, Command must be informed immediately.

This initial search is not fire attack or overhaul. Nothing but doors should be opened. Uncoordinated ventilation is a common factor resulting in firefighter fatalities. Doors should be closed after the search is complete in order to compartmentalize and confine the fire and other products of combustion.

The search for victims should continue until the area is completely searched or until fire conditions dictate otherwise. Crews working above the fire should perform a risk assessment as part of their size-up. The following questions should be considered:

- Are there savable victims in the area?
- Is it safe for the search team to be there?
- What are the priorities?

As the search progresses, it's important for the crew to stay focused on the task, to eliminate undue risk, and to communicate their position and findings regularly.

The Oriented Search

FOCUS OF THE ORIENTED PERSON

We have mentioned the oriented search several times in this module. We will now discuss the execution of this method in detail. The oriented search is based on focus. One member, called the "oriented person" or "oriented man" (usually the company officer), has as his or her sole focus the safety of the crew. This individual maintains constant awareness of three things:

- the location of the entire crew
- the fire conditions in the area
- the location and progress of other searchers

While the oriented person focuses on the safety of the crew, the searchers maintain their focus on one task—they search!

During the 1980s, tests conducted in blacked-out facepiece conditions and during actual fires revealed that searches were conducted more rapidly with an oriented search. Search teams got lost less frequently, if at all, and the firefighters conducting the search did a more thorough job. Additionally, the firefighters doing the search felt more at ease and more confident. The number of victims found was the same on a room-for-room basis, but more rooms were searched using the oriented search method in larger buildings.

Elements of a successful search

TRACKING SEARCHERS

There are a number of elements that must be present for a successful oriented search. The first is verbal communication. Constant dialogue between the searchers and the oriented person helps calm each searcher and track where each one is. If the searchers have a confidence in the oriented person's position and are confident that the oriented person is maintaining communication with the outside forces, then the searchers are free to focus on the search, and performance is improved.

COMMUNICATING FINDINGS

The searchers will also communicate freely about what they are finding. They will tell the oriented person what is in the room, including furniture, contents, bathrooms, closets, windows, etc. This report keeps the crew informed as to what types of rooms have been searched and what direction the team is traveling within the structure. Communicating the number of walls in a room helps everyone visualize the room and ensures that all areas are searched. It also provides an easy way to reverse direction and get back to the oriented point.

DIRECTION OF SEARCH

Coordinating the direction of the search is also important. The searcher will communicate what search pattern is being used. The most common search options are:

- left-handed
- right-handed
- left off hoseline (commercial)
- right off hoseline (commercial)

Search process

In an oriented search, both the oriented person and the searcher enter the structure. The oriented person leads the way and memorizes how they get in and how to get out. As the team leader, the oriented person locates the area to be searched. The searcher then enters the search area and communicates the number of walls and the direction of search. The search crew members maintain constant communication with each other as the search progresses in each room.

As the search team searches, the oriented person locates the next room to be searched. He or she checks the area to be searched for changing heat and smoke conditions. This person should also obtain information from other crews behind the search area to keep abreast of fire conditions on all six sides—north, south, east, west, up, and down. When the search team is finished, the oriented person leads the searchers to a new area to search, and the process is repeated.

In large commercial buildings, there are several precautions to take. In general, the oriented person stays with the "oriented point" except to take quick "sorties" to check the fire or to check searchers. The designated "oriented point" should be constant and nonmovable if at all possible.

Search Summary

Managing the search process is one of the most important tasks at any offensive fire. The search officer must coordinate the efforts of the search team while maintaining enhanced firefighter accountability and a secured means of escape from the hazard zone. In addition, the search team must maintain total focus on locating victims who are trapped by fire and smoke, in order to complete the search successfully. The oriented search is one method to improve the way firefighters conduct one of the most important job functions that they are sworn to carry out.

Ventilation Theory*

Throughout this course, ventilation is discussed in relation to smoke. Smoke is a convenient medium to use in illustrating air movement because of its visibility. However, the same principles apply to many toxic fumes, paint fumes, welding fumes, vehicle exhaust, dust, and other airborne contamination.

FUNCTION OF VENTILATION

Ventilation is the replacement of the products of combustion, heat, and fire gases with fresh air. On the fireground, ventilation is essential to the overall objective, in that it enables firefighters to complete their mission of rescue and extinguishment more rapidly and safely. Through a coordinated attack that includes the ventilation of the fire building, firefighters will work in an environment that is more tenable for themselves and for any trapped occupants in the building. Ventilation reduces the conditions of backdraft and flashover in addition to increasing visibility for other fireground operations.

THE SOONER VENTILATION OCCURS, THE GREATER THE REDUCTION IN PROPERTY DAMAGE.

Fireground commanders are now realizing the benefits of ventilation through the efforts of many years of research by manufacturers, fire departments, and insurance companies, as well as through their own experiences. Research has shown that the sooner the ventilation process occurs in the overall fireground plan, the greater the reduction in property damage, and the greater the reduction in manpower and equipment needs. As fuels have changed over the past several years, fire gases have become considerably more toxic, and the ventilation objective has become more and more important. The increase in the toxic content of fire gases has had a dramatic effect on the way we attack fires.

* The material that follows is adapted from the *Smoke Ventilation Training Manual*, published by Super Vacuum Manufacturing Company, Inc., P.O. Box 87, Loveland, CO 80539-0087. Used by permission.

NATURAL AND MECHANICAL VENTILATION

Ventilation can be achieved through natural or mechanical means. Natural ventilation includes opening windows and allowing the wind to provide fresh air to the fire building environment. It also includes self-ventilation of the fire when it burns through the roof, thereby allowing the trapped fire gases to escape through the roof opening. Mechanical ventilation includes using HVAC systems built into buildings, using fog streams out of windows to create air movement, and negative or positive pressure ventilation through the use of specialized fire service fans. This section will concentrate on the use of specialized fire industry fans to create positive and negative pressure ventilation.

Why Ventilate?

STAGES OF FIRE

As discussed in Module 1, during the incipient stage of fire (the early stage), the oxygen content of the environment is not changed dramatically. The fire is, however, adding carbon dioxide (CO_2), carbon monoxide (CO), sulfur dioxide (SO_2), water vapor (H_2O), and other gases to the atmosphere. During this phase of fire, flame temperatures reach 1,000°F, while the temperature in the room may only change a few degrees. The products of combustion—CO_2, CO, SO_2, and H_2O—all have oxygen in their makeup (O or O_2). This means that oxygen is being used during the combustion process. If a continuous supply of oxygen is not available, the fire will go out (smoldering or decay phase).

The second phase of fire is the free-burning phase. This phase cannot occur without the continued supply of oxygen. As air is introduced to the base of the flame, heat and fire gases will rise until stopped by the ceiling usually, or until they cool to equal ambient temperature. These superheated gases may reach temperatures of 1,200°–1,500°F. The inhalation of these fire gases can cause instant death.

In the smoldering or third phase of fire, burning is reduced due to the lack of oxygen. The greatest smoke generation occurs during this phase. So much smoke may be generated that it pressurizes the structure with smoke. Smoke may be seen from the exterior of the building. Smoke and fire gases may still be above 1,000°F.

SMOKE AND HOT FIRE GASES KILL MORE PEOPLE AND CAUSE MORE DAMAGE THAN FLAMES.

During a fire, different areas of the structure may be at different stages of combustion. Smoke and hot fire gases kill more people and cause more damage than flames. In addition, these products of combustion make it more difficult to perform rescue and extinguishment operations because of low visibility and physical stress placed on personnel.

Smoke and Fire Gases

Smoke is the mixture of the products of combustion with air and dust. These products include, but are not limited to, carbon, organic acids, aldehydes, and tar. Many products of combustion are heavier than air; however, in the heated environment of the fire, they expand and rise. As the room cools, these lethal fire gases become denser and fall to the floor level. For this reason, ventilation and the use of breathing apparatus must continue during the overhaul phase of operations. The use of breathing apparatus does not preclude the use of ventilation techniques; many products of combustion may be absorbed through the skin, especially in the areas of the firefighter's neck and wrists. Firefighters assigned to busy companies have an additional risk, in that fire gases are cumulative in the body. Conditions that affect the level of danger include the following:

CONDITIONS AFFECTING LEVEL OF DANGER

- level of oxygen in the area
- temperature of the fire gases

- products involved in the fire
- duration of exposure
- physical condition of the firefighter

Ventilation can affect all but the physical condition of firefighters, thereby reducing the risk considerably. Firefighters exposed to the products of combustion should be given 100% oxygen therapy. Again, ventilation reduces the risk. The long-term effects of exposure to fire gases are still not known. However, we do know that carcinogens are produced during the combustion of some products. In addition to the carcinogens, immunosuppressants are produced during fires. These suppressants inhibit the immune system of the body from protecting itself.

LONG-TERM EFFECTS OF EXPOSURE

Flashover

Flashover occurs when the contents of a room or area reach their ignition temperatures almost simultaneously. Flashover may appear as a wave, because as one space ignites, the increase in temperature causes the next space to reach its ignition temperature. Ventilation procedures can reduce the flashover risk by reducing the temperature in the entire area.

USE OF VENTILATION FOR FLASHOVER AND BACKDRAFT CONDITIONS

Backdraft

A backdraft situation exists when the environment should be in a flashover; however, there is not sufficient oxygen to support combustion. When oxygen is introduced into this environment, an explosion occurs as the fire spreads.

Ventilation is required in this case. It is important to understand the entire ventilation process and carry it out carefully under backdraft situations. As mentioned in Module 1, the following are some signs of backdraft:

- puffing smoke outside the building or area
- dense, black smoke turning gray to yellow
- little or no flame
- smoke-stained windows
- smoke leaving and then reentering the building
- excessive heat with no visible fire

Ventilation Tactics

Fire attack and/or rescue operations should occur around the same time as ventilation operations or soon thereafter. Coordination of the fire attack and ventilation operations is essential, as ventilation operations dramatically affect the outcome of the situation. Ventilation can be destroyed by attack crews who are not acting in concert with ventilation crews. Also, attack operations not in conjunction with ventilation operations may not reach their objective, either the fire or the rescue. After completing ventilation operations, ventilation crews should return to or communicate with the command post for reassignment.

Coordination of the fire attack and ventilation operations is essential.

Protection lines for the ventilation crew should not be used for firefighting lines unless specifically directed by the command post. Attack crews could be in jeopardy if ventilation crews fight the fire from the wrong vantage point. The tactical objective of the ventilation crew will determine where and how ventilation is done. For rescue, generally, positive pressure is needed between any trapped occupants and the fire and between trapped occupants and the exit. Once the rescue mission is completed, ventilation equipment may need to be repositioned for attack or overhaul operations. During attack operations, positive pressure is needed between attack crews and the fire, with the exhaust opening on the opposite side of the attack.

Whenever possible, ventilation crews should take advantage of the natural tendency of smoke and fire gases to rise.

Whenever possible, ventilation crews should take advantage of the natural tendency of smoke and fire gases to rise. While a potentially dangerous operation, some situations call for firefighters to make their way to the roof and cut it open. This will release superheated fire gases, smoke, and sometimes fire through the hole. A protection line is required. Never put the protection line on or in a vent hole unless it is for the direct protection of the ventilation crew. Once a ventilation hole has been cut, the ventilation crew should retreat to the ground for reassignment.

On a smaller scale, firefighters can take advantage of smoke's tendency to rise by placing fans at the highest point in the exhaust opening. This opening may be a window or a door, or it may be an opening created during the ventilation process. During positive pressure operations, firefighters should ventilate the bottom floor first and work their way to the top.

Mechanical Ventilation

WHEN TO USE MECHANICAL VENTILATION

Mechanical ventilation is required whenever natural ventilation is even partially unsatisfactory. Mechanical ventilation procedures should begin immediately to assist in rescue or attack operations, or when it is felt that the natural process is too slow, causing excessive damage to property. Mechanical ventilation should be used to maintain egress routes for occupants and firefighters. It should also be used to augment natural ventilation or when smoke is below grade. Generally speaking, if ventilation is applied correctly, you cannot overventilate. More often than not, the fire service underventilates during fire operations due to lack of training, severe manpower limitations, poor equipment, and/or a misunderstanding of the value of ventilation procedures.

HOW MECHANICAL VENTILATION WORKS

Mechanical ventilation is made possible by creating positive and negative areas and directing air currents, with the principle that air will travel from a positive or neutral area of pressure to an area of negative pressure. Using mechanical force, we can create negative or positive pressure on the fireground in one of two ways:

- specially designed smoke ventilators
- fog streams

These aids to ventilation will be discussed below.

Pressure Gradients

When smoke and fire gases are on the side of the fan where the pressure is positive, that is called positive pressure ventilation (PPV). When smoke and fire gases are on the negative side of the fan, that is called negative pressure ventilation (NPV). All electric ventilation fans may be used either in the positive pressure operation or in the negative pressure operation. Using any gasoline powered fan in the negative pressure mode is not recommended due to the smoke and gases being pulled into the carburetor. Firefighters should know how to use both positive and negative pressure procedures, as different situations call for different operations. There are no rules for determining when to use positive pressure and when to use negative pressure ventilation. In many situations, both positive and negative pressure procedures will be in operation at the same time.

Figure 5-1 presents a flowchart that can be used during scene size-up to assist with ventilation decisions. Figure 5-2 summarizes the requirements for effective mechanical ventilation.

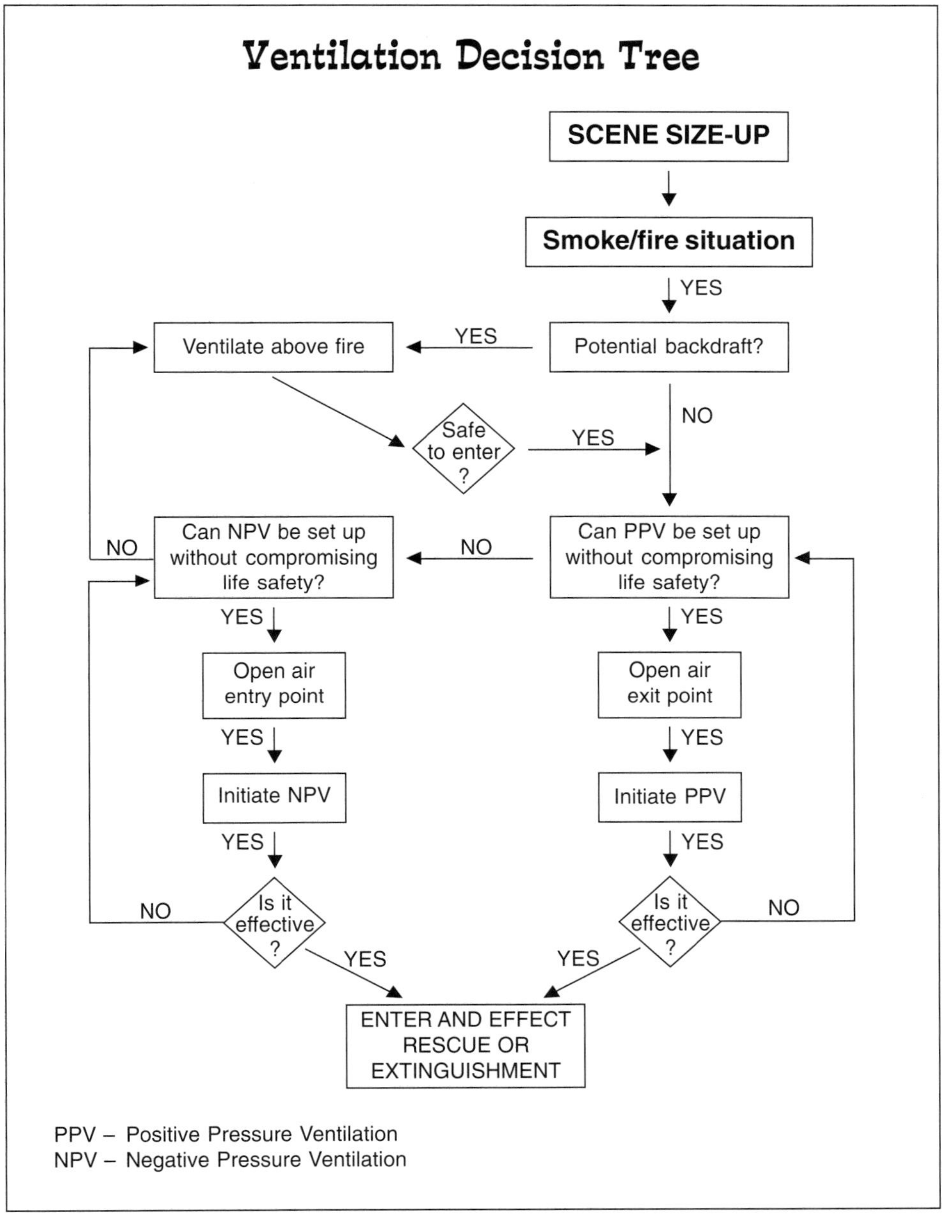

Figure 5-1: Ventilation decision tree for a smoke/fire situation.

Requirements for Effective Mechanical Ventilation

1. Use good, reliable, portable equipment.
2. Be sure to use large enough fans or enough fans to perform the necessary task.
3. Provide proper training in techniques to acquire maximum benefit from the equipment.
4. Practice good prefire planning that includes ventilation options.
5. Take advantage of smoke's natural tendency to rise.
6. Initiate ventilation as soon as tactically possible.
7. Include ventilation as part of standard operating procedures (SOPs).
8. During negative ventilation operations, place fan as close to smoke source as possible.
9. Use combinations of fans to achieve the best performance, either combinations of positive and negative pressure, or multiple fans for additional volume.
10. Use prevailing winds as much as possible.
11. Keep air paths clear. Take advantage of air's natural tendency to follow the path of least resistance.
12. Remove screens and other obstacles. Maintain the closure integrity so the window or door can be closed.
13. Coordinate the ventilation operation with rescue and attack operations.
14. Maintain control of replacement air and/or exhaust openings.
15. When ventilating more than one floor, start at the bottom and work toward the top.
16. Never use gasoline powered fans in smoke atmospheres.
17. Use only fans with explosion-proof motors inside questionable atmospheres.
18. Never place hoselines into ventilation openings.
19. Make sure that created drafts do not close doors or windows.
20. Maintain ventilation operations during overhaul.
21. First open, then pressurize, maintaining control of air movement.

Figure 5-2

Churning

During mechanical ventilation, a division must be maintained between the positive side of the equipment and the negative/neutral side. If a separation is not made, air will flow from the positive side back around to the negative side, making the operation less effective. This phenomenon is called churning.

Using Combinations of Fans

Using more than one fan increases air input and speeds up the ventilation operation. Multiple fans can be positioned at a single inlet (Figure 5-3), using more than one electrical outlet, or several fans can be stacked. Combinations can also include both positive and negative pressure operations (Figures 5-4, 5-5, and 5-6). It is not recommended that gasoline fans be used indoors. Electric fans may be used either indoors or outdoors.

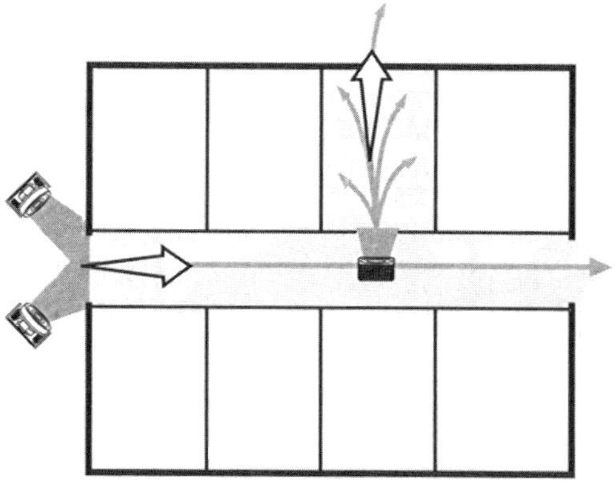

Figure 5-3: Multiple fans at inlet, with positive pressure fan changing the direction of the air current.

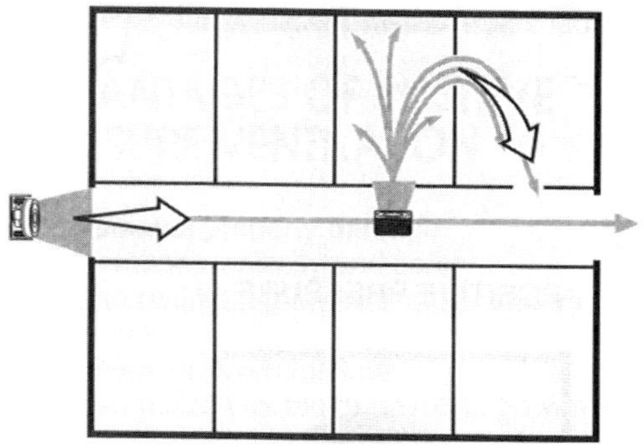

Figure 5-4: Single fan at inlet, with positive pressure fan changing the direction of the air current.

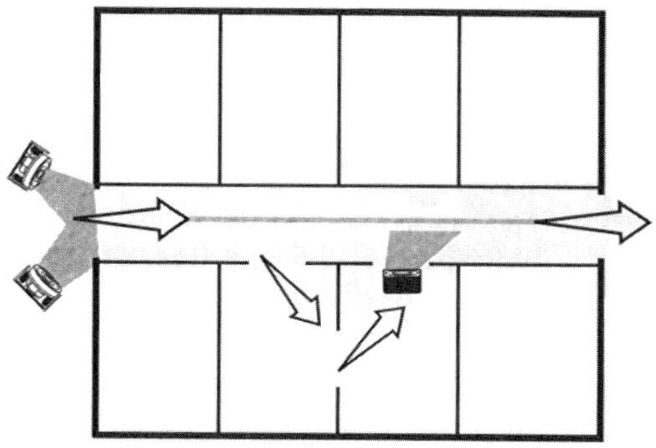

Figure 5-5: Multiple fans at inlet, with negative pressure fan changing the direction of the air current.

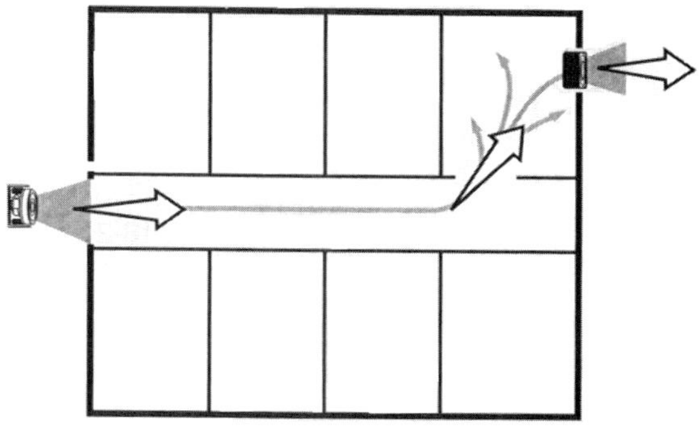

Figure 5-6: Single fan at inlet with negative pressure fan at outlet.

When combinations of fans are used, it is very important to allow sufficient openings for air to flow to a fan in negative pressure operation and from a fan used in positive pressure operation. If openings are too small, the efficiency of the fan will be reduced.

The older, series method of using fans (Figure 5-7) is not as effective as parallel methods. This is similar to pumps moving water, where parallel pumps will move more water than series pumps.

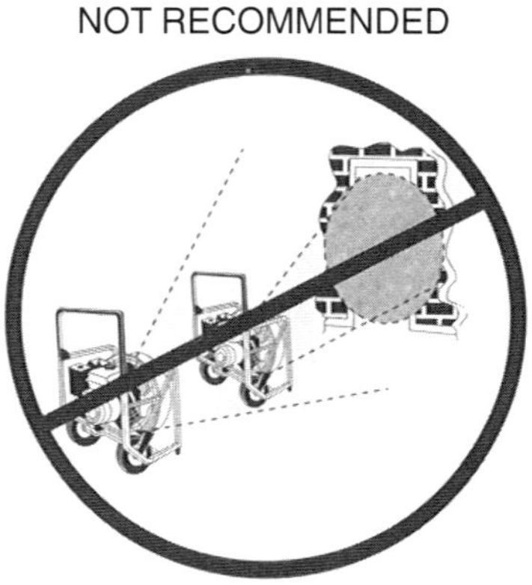

Figure 5-7: Series method of fan placement.

The V-attack procedure (Figure 5-8) will ventilate faster and more effectively than series PPV while allowing straight-in access. Place fans at a 45° angle to the door, with one aimed high and the other aimed low.

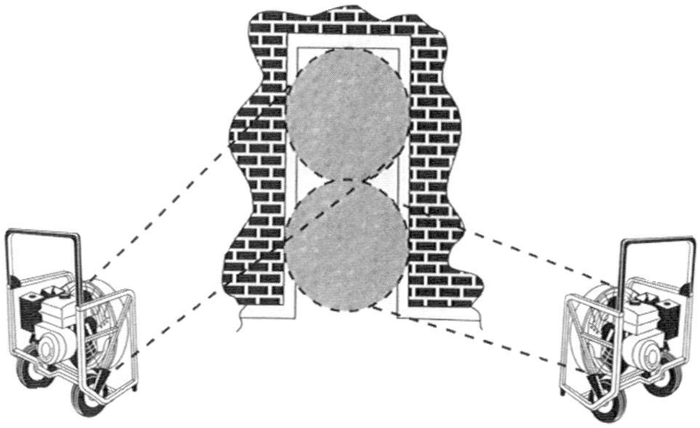

Figure 5-8: V-attack method of fan placement.

Fanning a Fire

It is important to realize the relative speed of air in ventilation situations. This can aid the ventilation team in determining the number of fans and ventilation method required. The average windspeed next to a fan is 60 mph, in a doorway the airspeed is about 10 mph, and inside a room it is about 2 mph (Figures 5-9, 5-10, and 5-11).

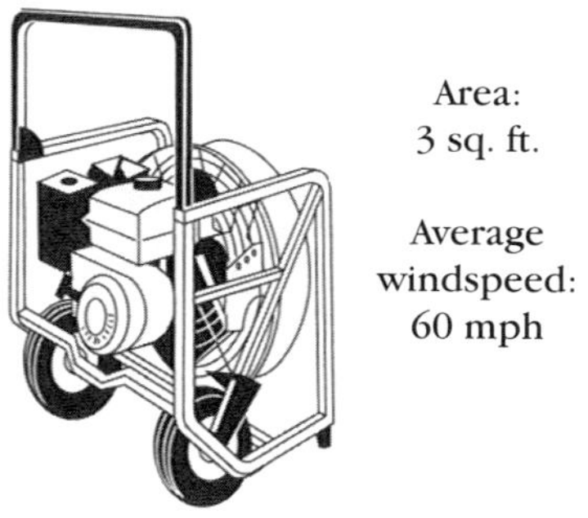

Figure 5-9: Windspeed at fan.

Figure 5-10: Windspeed at a door.

Figure 5-11: Windspeed in a typical room.

Using Wind Direction

One of the most important requirements of successful ventilation is knowing how to use the prevailing wind to your advantage. Once the windspeed and direction are determined, the ventilation team can begin their setup, using either positive or negative pressure ventilation to clear the smoke most effectively (Figures 5-12 and 5-13).

One of the most important requirements of successful ventilation is knowing how to use the prevailing wind to your advantage.

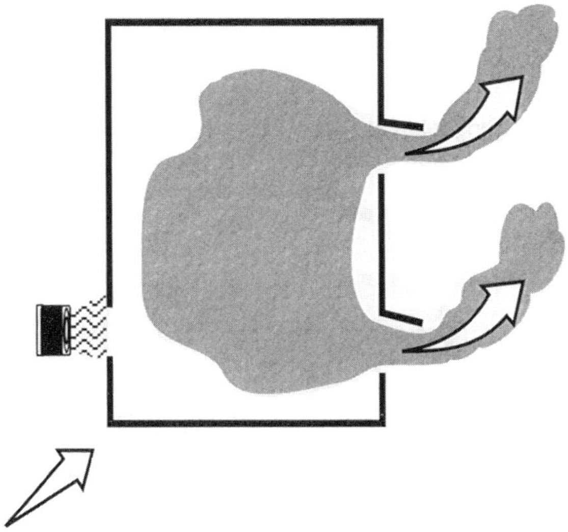

Figure 5-12: Using wind direction with positive pressure ventilation.

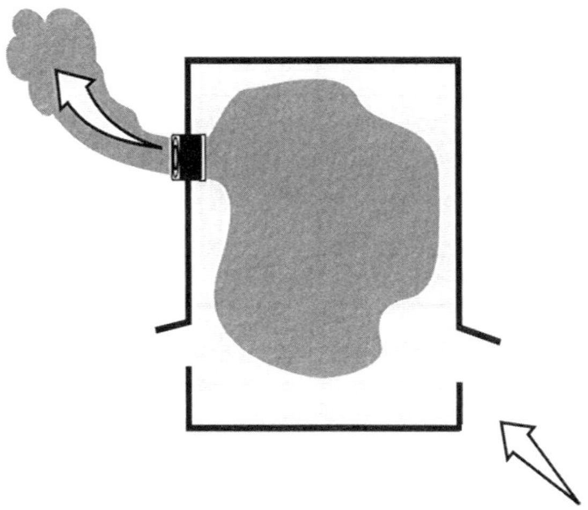

Figure 5-13: Using wind direction with negative pressure ventilation.

Positive Pressure Ventilation

Positive pressure ventilation (PPV) is defined as the introduction of fresh air into an area through a fan in order to increase the pressure in that area. When the pressure is increased, the air will move to an area of lower pressure. This follows the law that air will always move from an area of higher pressure to an area of lower pressure. During this movement, the air will always follow the path of least resistance.

WHERE TO INITIATE POSITIVE PRESSURE VENTILATION

Firefighters can take advantage of the positive pressure laws in the area of smoke removal. By pressurizing a structure and by using the selective process of opening and closing doors and windows, firefighters can direct the movement of smoke and fire gases faster than ever before. Because positive pressure ventilation is initiated from outside the structure, firefighters can implement ventilation operations immediately, without entering the hostile environment. This will help cool the space, provide oxygen to victims, and increase visibility for victims and firefighters. As with all ventilation practices, firefighters performing ventilation should know where the fire is before they begin, as the ventilation process has the potential of moving fire and fire gases. Generally speaking, positive pressure should be initiated between firefighters and the fire or between the fire and trapped occupants. The exhaust port should be on the opposite side of the firefighters or the occupants. Pressurization should occur after opening any exhaust port.

Advantages of PPV

- Reduces property damage.
- Increases visibility and safety.
- Can maintain primary and secondary egress routes.
- Reduces overhaul time.
- Can utilize electric or gasoline powered fans.
- Does not require explosion-proof motors.
- Can be initiated from the exterior of the building.
- Is effective on all structures when doors or windows are maintained.

- Does not require smoke and fire gases to pass through the fan.
- Is quicker than negative ventilation.
- Keeps the doorway clear of equipment.
- Increases the air velocity within a building.
- Reduces equipment cleaning time.
- Works well in large areas.

Fan Placement

CREATING A VENTURI EFFECT

Working from the outside of the doorway, place the fan far enough away from the doorway so as to fill the doorway with a cone of air coming from the fan (Figure 5-14). This will create a venturi effect with the air going in the door, increasing the fan's effectiveness. In large door openings, place the fan slightly farther back or use multiple ejectors.

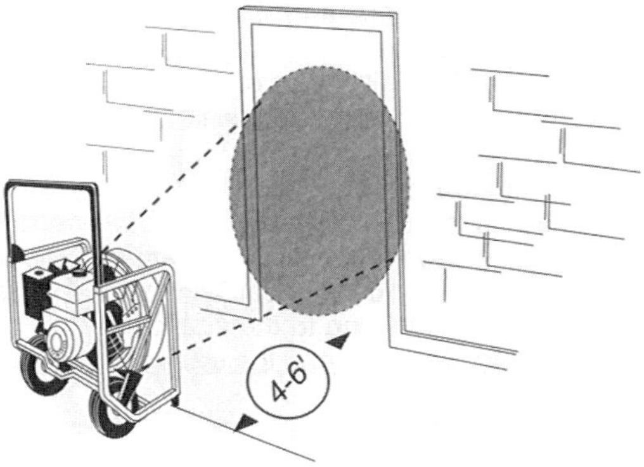

Figure 5-14: Correct placement of one fan for positive pressure application.

Parallel fans will increase volume greatly, reducing ventilation time and increasing ventilation effectiveness. For oversized doors, place fans side by side to overlap their cones of air.

Positioned as described above, the fan not only moves the air passing through the propeller, but also draws air into the airstream on the discharge side. Efficiency increases by 20%–100% of rated capacity. The actual increase varies according to the size of the fan, the size of the door opening, and the size of the exhaust opening.

The smaller the fan, the lower the increase in capacity, and vice versa. This is mainly because of the difference in velocity and the concentrated reach of the airstream. Positioning either an 18" or a 24" ventilation fan 4–6 feet outside a door measuring 30" x 84" will cause the volume of air exhausted to increase about the same for either fan (30%–40%).

The design of the V-attack is such that the door is sealed from top to bottom, with the units at 45° angles. This leaves a straight shot for attack lines (Figure 5-15).

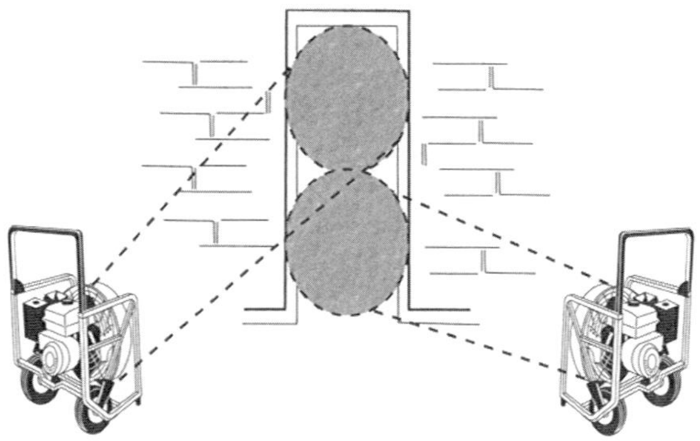

Figure 5-15: Correct placement of two fans for positive pressure application.

The design of the saddle stack is such that all the air seals the door (Figure 5-16).

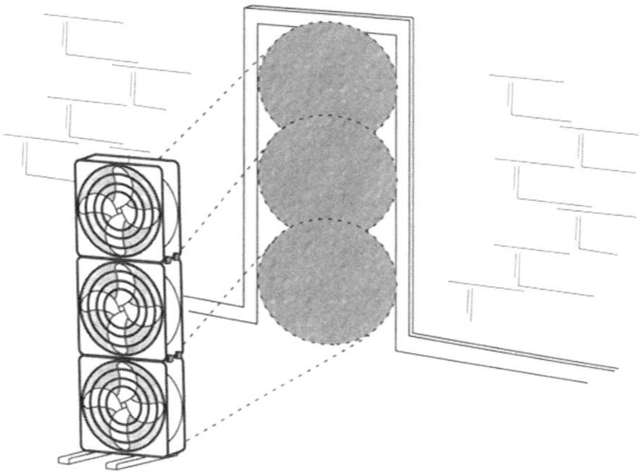

Figure 5-16: Saddle stack placement of fans.

Use of Doors

Because fans are not located directly in the doorway for positive pressure operations, the ingress/egress route is not blocked by the fan. Firefighters entering the building should be careful not to stand in the doorway, as that will obstruct the incoming air. Similarly, personnel must be careful not to block the air supply by standing or stooping in back of the fan (Figure 5-17).

Figure 5-17: Do not block the air supply.

Streamers can be hung in a doorway to indicate that it is a positive pressure ventilation inlet (Figure 5-18).

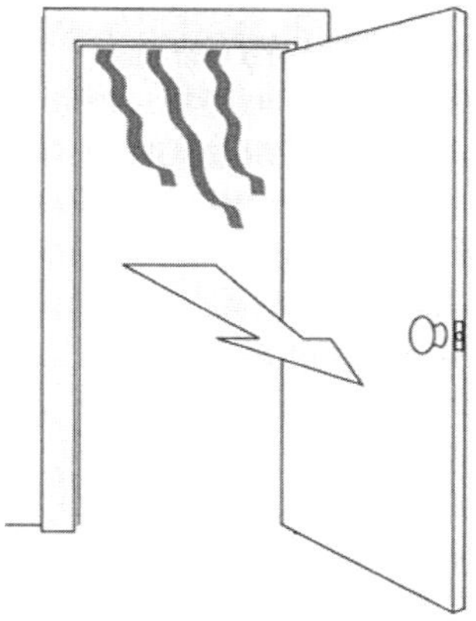

Figure 5-18: Using streamers in a doorway.

When using a garage door for ventilation, open it only partway. Then using two fans, seal the doorway with the cones of air (Figure 5-19).

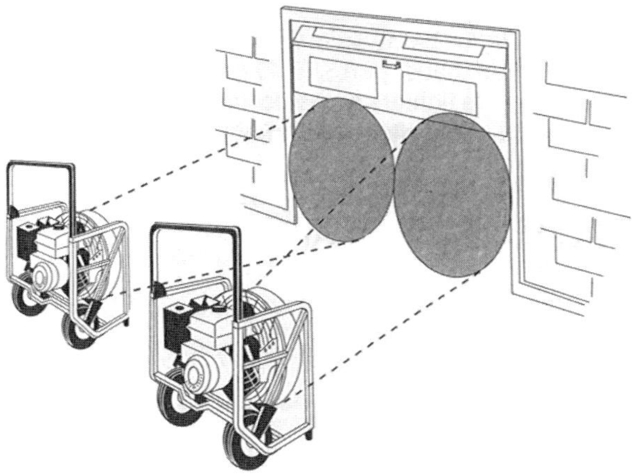

Figure 5-19: Using a garage door for ventilation.

Working With Positive Pressure

All ventilation operations should be under the direction of the incident commander. This can be direct supervision, through a company officer such as a truck company officer, or through a ventilation section leader. Ventilation operations are as critical to the overall operation as are other procedures. If the ventilation operation is neglected, the overall operation suffers.

If the ventilation operation is neglected, the overall operation suffers.

Once a crew has been assigned to perform the ventilation operation, the members must identify the location of the fire and the mission of the team. For example, if there is a backdraft condition, the crew should open up above the fire to exhaust superheated fire gases and products of combustion. If a rescue is to be made, the crew needs to place a positive pressure zone between the trapped persons and the fire and another positive pressure zone between the rescue crew and the fire. When making a direct attack, the ventilation crew needs to open up on the opposite side of the attack crew, then place a pressure zone between the attack crew and the fire. To initiate positive pressure operations, either gas powered or electric powered fans may be used. Generally speaking, a larger fan will produce faster results. However, always compare fans by their rated output, not by their size in inches.

Care should be taken not to destroy the integrity of windows or doors so that they can be opened and closed to change the direction of air flow.

Because positive pressure requires the select opening and closing of air exit points, care should be taken not to destroy the integrity of windows or doors so that they can be opened and closed to change the direction of air flow. This is contrary to many departments' standard practice and will require new awareness and training. Also, firefighters using several doors as ingress and egress routes can reduce the effectiveness of the ventilation effort; again, training and coordination are important (Figures 5-20 and 5-21). If windows are broken, doors can be used to compartmentalize rooms.

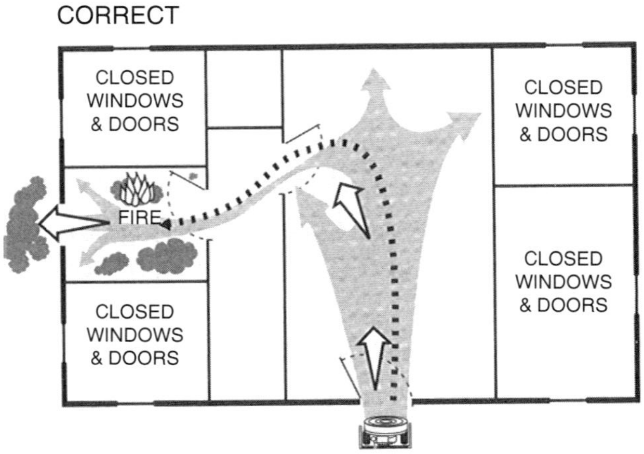

Figure 5-20: The attack crew should work in concert with ventilation.

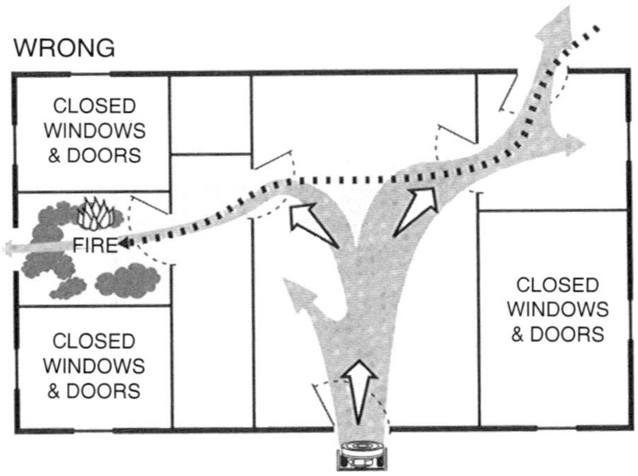

Figure 5-21: An uncoordinated attack will work against both attack and ventilation.

Negative Pressure Ventilation

NPV has been the industry standard for many years, and it remains the most versatile tool in ventilation.

Negative pressure ventilation (NPV) is the movement of smoke through a fan from the negative side first. NPV has been the industry standard for many years, and it remains the most versatile tool in ventilation. It is the only recommended method of ventilation in situations involving hospitals, nursing homes, cave-ins, and interior rooms. Electric fans do the majority of negative pressure work, since they are not affected by the ambient environment, do not produce carbon monoxide, and can be used in any position.

Advantages of NPV Using Electric Smoke Ejectors

- Reduces property damage.
- Increases visibility and safety.
- Can maintain primary and secondary egress routes.
- Reduces overhaul time.
- Can utilize electric fans.
- Can utilize explosion-proof motors.
- Does not require a great deal of coordination with other operations.
- Is not upset easily.
- Requires little training.
- Can be used in conjunction with flexible duct equipment.
- Works well in small-area situations such as a single room.
- Works well in the removal of heavier-than-air gases.
- Works well from aerial equipment from the exterior of high-rise rooms.
- Works when opening integrity is not maintained.
- Electric smoke ejectors do not produce carbon monoxide.
- Can be used in cave-ins.
- Can be used to cool firefighters in staging areas.

Window Placements

Two principal kinds of windows are encountered—the double-hung window and the casement-type window.

Double-hung window

1. Raise lower window.
2. Place the ventilation fan unit on the window sill and pull the window down onto the handles.
3. Secure both hooks of the fan hanger unit into the bottom sill of the window.
4. Wrap the cable of hooks around the handle to obtain a length which will hold the smoke ejector securely on the lower sill of the window.

Properly installed, the fan unit performs unattended and ejects smoke at maximum efficiency (Figure 5-22). NOTE: Be sure the unit is correctly placed. To direct smoke, the fan blade must be on the inside, forcing air over the motor to the outside. Reverse the position to pull in fresh air.

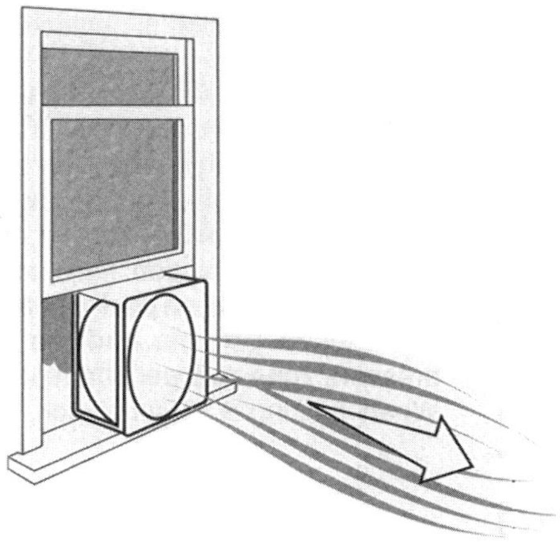

Figure 5-22: Fan placement in a double-hung window.

Casement window

1. Roll the window open and place one hanger hook over the top window hinge.
2. Fasten the second hook into the mullion of the window to steady the unit.
3. Rest the unit on the metal edge of the window to prevent damage.

This placement (Figure 5-23) is of tremendous value in venting the common mattress fire or similar one-room fires.

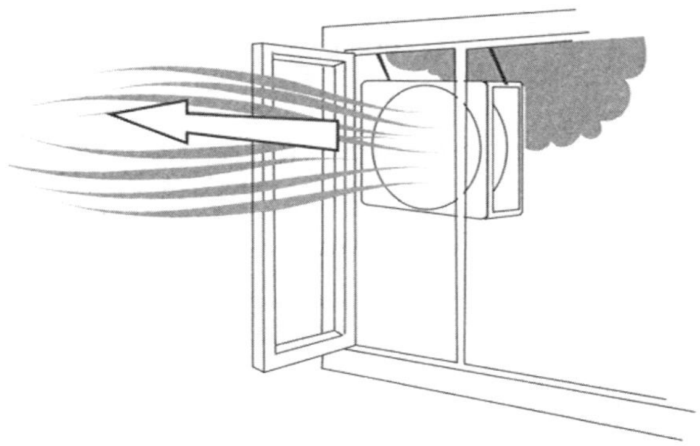

Figure 5-23: Fan placement in a casement window.

Negative Pressure Door Placement

With the fan on the smoke side of the doorway, place a fan or fans far enough away from the doorway so as to cover the doorway with a cone of air coming from the fans (Figure 5-24).

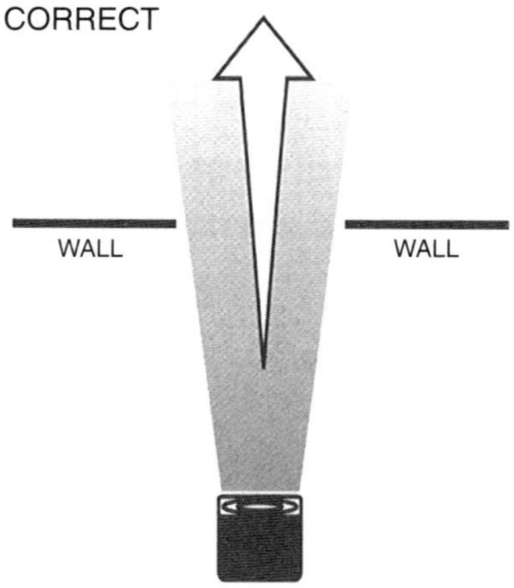

Figure 5-24: Proper placement of a fan outside a doorway.

Placing the fan too close to the doorway can produce recirculation (Figure 5-25).

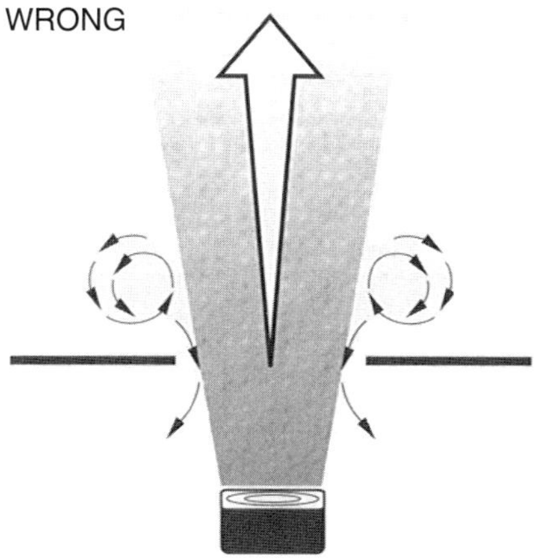

Figure 5-25:
Fan placed too close to opening, resulting in recirculation.

If the fan is placed too far away from the opening, its efficiency will be reduced (Figure 5-26).

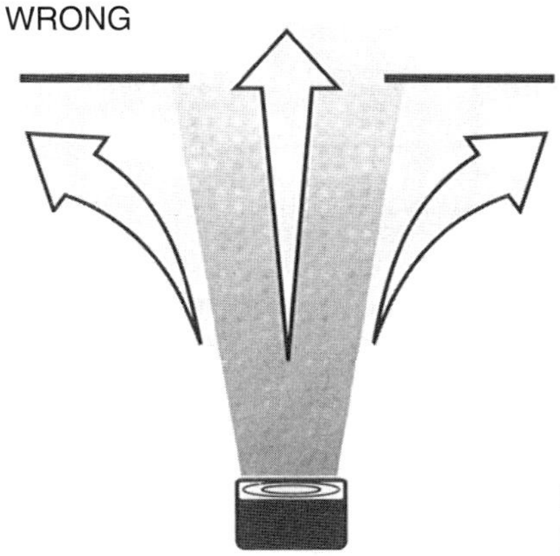

Figure 5-26:
Fan placed too far from opening.

When the fan is positioned properly, the cone of air produced will create a venturi effect with the air going out the door, increasing the fan's effectiveness by as much as 100% (Figure 5-27).

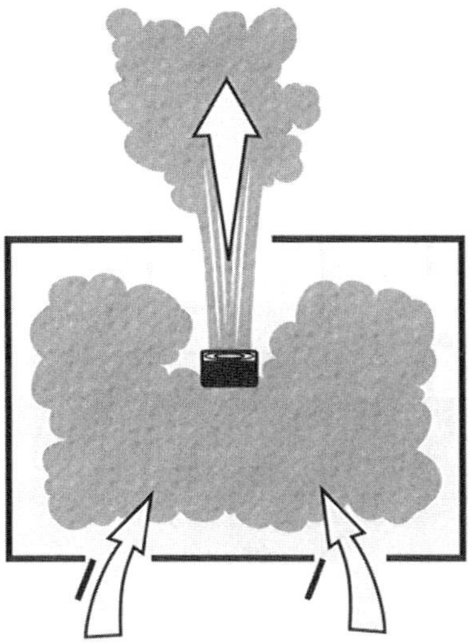

Figure 5-27:
Typical placement for efficient venturi effect (air inlet approximately 2 times air outlet).

Stacking fans will increase volume greatly, reducing ventilation time and increasing ventilation effectiveness. For oversized doors, place fans side by side to overlap their cones of air. An alternate method of using two fans in a doorway is to place one fan on the door, and a second fan away on the floor.

The following guidelines can be used as an aid in determining the correct placement of ventilation fans based on the sizes of the door and the fan.

GUIDELINES FOR FAN PLACEMENT

- A 16" ventilation fan exhausting through a 36" door should be positioned 6–8 feet from the opening.
- A 20" ventilation fan exhausting through a 48" door should be positioned 8–10 feet from the opening.
- A 24" ventilation fan exhausting through a 48"–60" door should be positioned 6–10 feet from the opening.
- In large door openings, place the fan slightly farther back or use multiple fans.

Positioned in this manner, the ejector not only moves the air passing through the propeller, but also draws air into the airstream on the discharge side. Efficiency increases by 20%–100% of rated capacity. The actual increase varies according to the size of the fan, the size of the door opening through which it is exhausting, and the size of the opening providing replacement or make-up air.

The smaller the fan, the lower the increase in capacity, and vice versa. This is mainly because of the difference in velocity and the concentrated reach of the airstream. By positioning either a 16" or a 24" ventilation fan 8–10 feet inside a door measuring 30" x 84", the volume of air exhausted will increase about the same for either ejector (30%–40%).

Working With NPV

A few of the situations requiring the use of NPV are vaults, closets, or closed rooms (Figure 5-28), attics (Figure 5-29), and internal rooms without outside access (Figure 5-30).

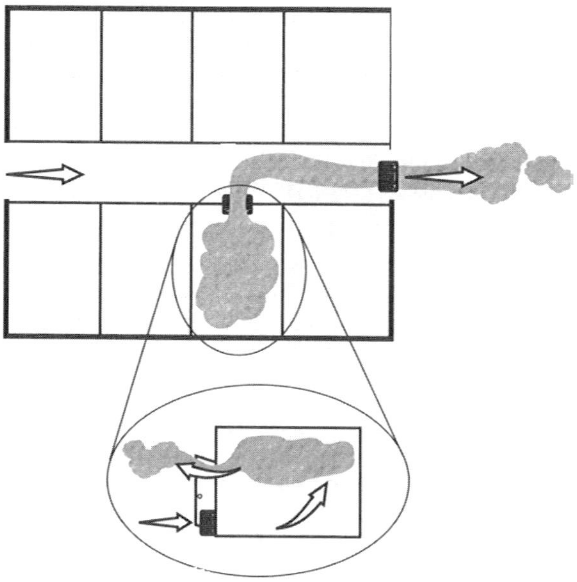

Figure 5-28: Venting vaults, closets, or closed rooms.

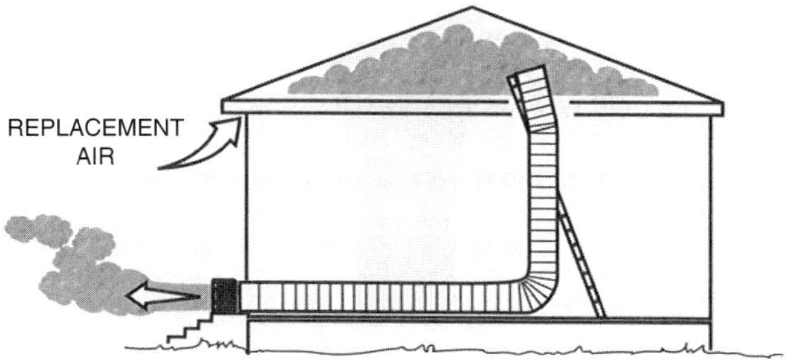

Figure 5-29: Venting an attic.

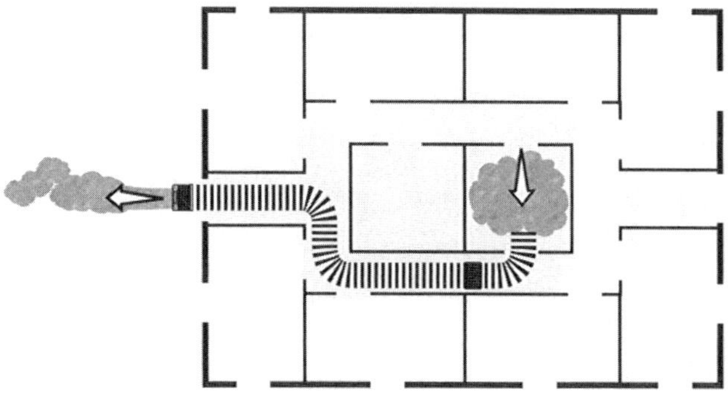

Figure 5-30: Exhausting internal rooms without outside access.

In order for fans to be used effectively, a path must be provided for make-up air to flow in and replace the smoke being pulled out through the fan. The sequence in Figure 5-31 shows how this may be applied to selective ventilation of an enclosure. NOTE: If windows have been broken, doors can be used to close off areas.

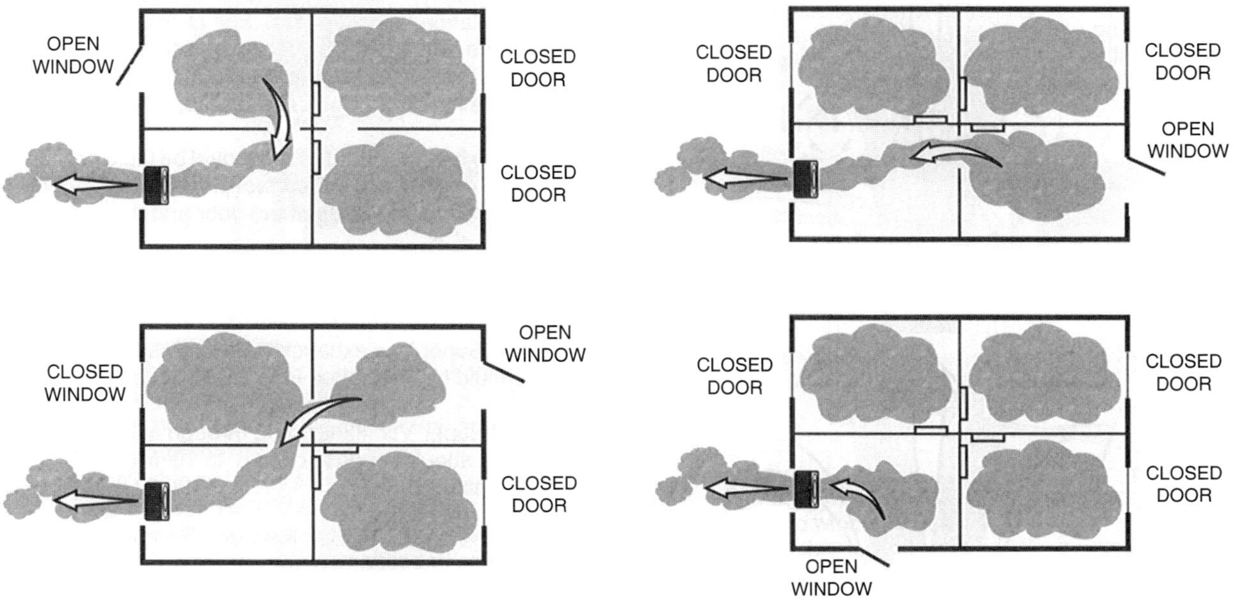

Figure 5-31: Selective ventilation of an enclosure.

Confined Space Emergencies

Rescue problems in confined spaces are usually a result of asphyxiation or due to falls. Unfortunately, many people have been seriously injured or killed in efforts to make confined space rescues without ensuring that the hazardous conditions have been removed. Confined space ventilation (CSV), which involves exhausting internal atmospheres and replacing those atmospheres with fresh air, is required before entry. Figure 5-32 presents a ventilation decision tree to use in confined space atmospheric emergencies.

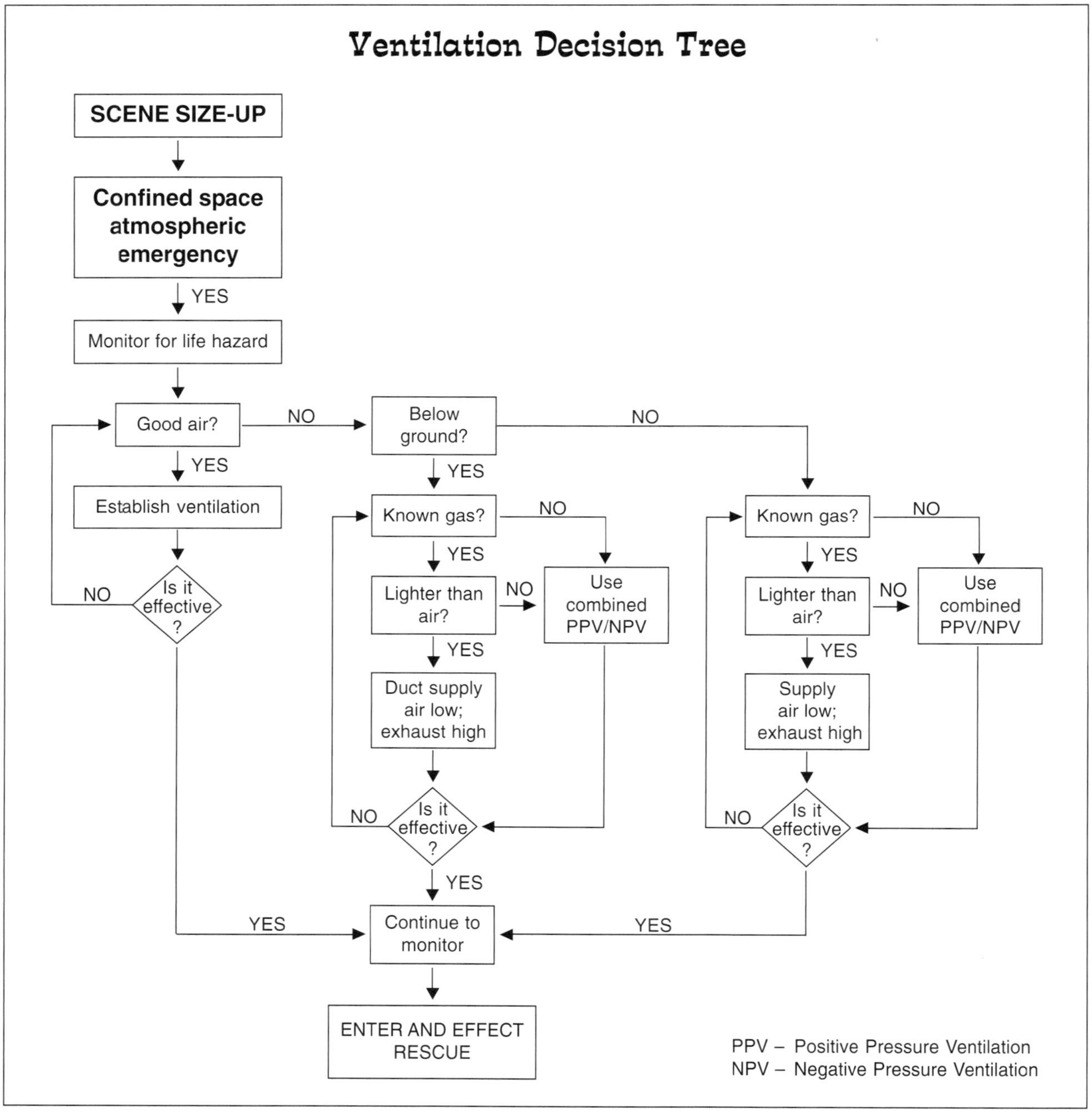

Figure 5-32: Ventilation decision tree for a confined space atmospheric emergency.

Extreme care must be taken when using gas powered blowers to provide air to confined spaces, because carbon monoxide is produced by the gasoline engine. A specialized electric fan used properly is the best ventilation tool for confined spaces.

A specialized electric fan used properly is the best ventilation tool for confined spaces.

In most cases, confined space ventilation can be set up without entering the space. This can greatly increase the chance of survival for victims in the space, without endangering the life of rescue personnel.

Safety Precautions for Confined Spaces

When using ventilation fans to ventilate confined spaces, always remember the following points:

- Never enter an area without monitoring the atmosphere.
- Air must be able to flow into and out of an area for effective ventilation.
- When explosive fumes are present, always pressurize an area by blowing into it with a fan.
- No fan is explosion proof. The fan motor may be, but any rotating device (blade) that is in close contact with a stationary device (housing) can cause a spark, either by static discharge or through a small metallic item entering the fan guards.
- Do not use a manhole adapter if the manhole is the only method of ingress/egress.
- The vapor density of the gases being ventilated must be known.
- Liquids in confined spaces should be completely removed before ventilation, if possible.
- If there is a sparking hazard, the fan or fan/duct combination must be grounded.

Confined Space Ventilation Techniques

Figures 5-33 through 5-38 illustrate ventilation techniques for various confined space situations.

In confined spaces containing heavier-than-air gases, it may be necessary to use more than one fan to ventilate properly. A fan with an explosion-proof motor and spiral duct can be used to pull a negative pressure into the area low enough to remove the fumes. Ideally, another more powerful fan should be used to supply clean air to the area. This second fan pressurizes the space so contaminants are not pulled into the space from unknown sources (see Figure 5-37).

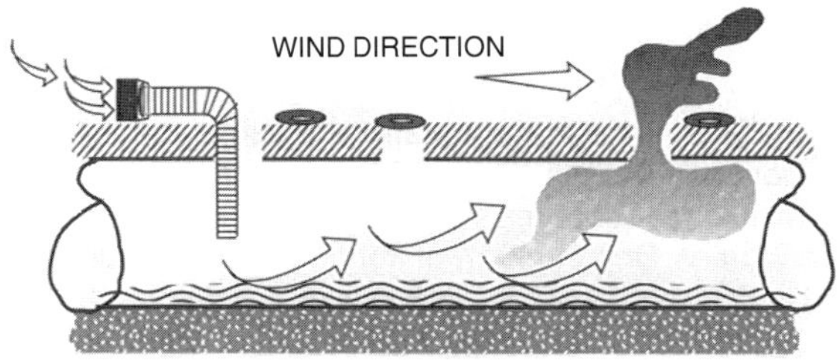

Figure 5-33: Using a duct to ventilate a confined space with lighter-than-air gases and more than one opening.

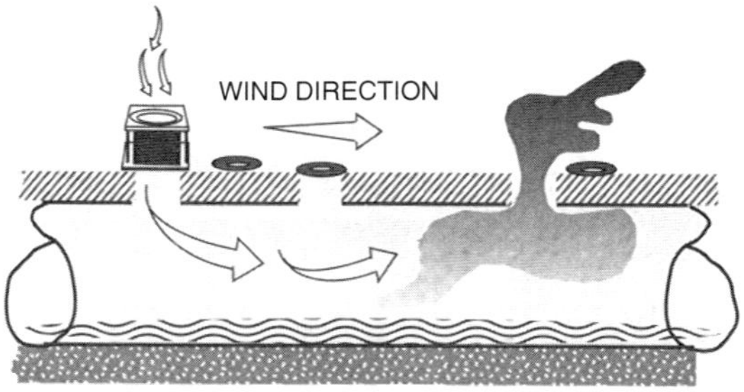

Figure 5-34: Using a fan to ventilate a confined space with lighter-than-air gases and more than one opening.

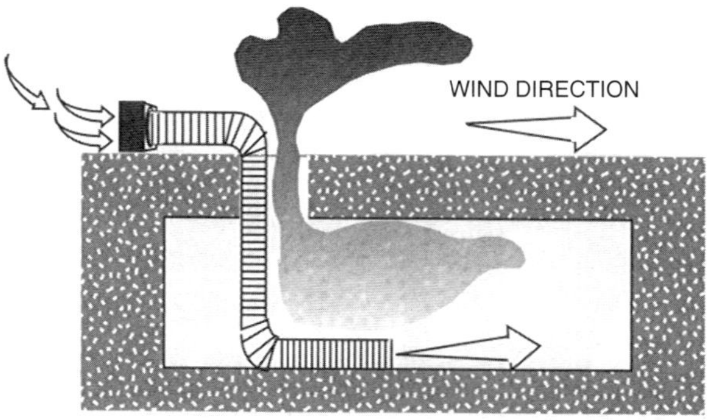

Figure 5-35: Ventilating a confined space with lighter-than-air gases and only one opening.

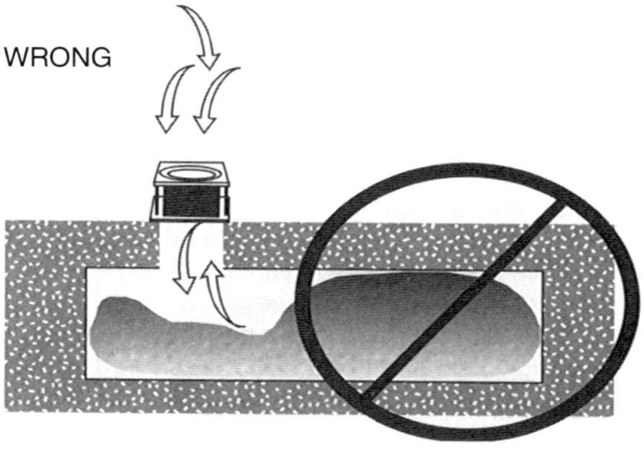

Figure 5-36: Improper ventilation of a confined space with lighter-than-air gases and only one opening.

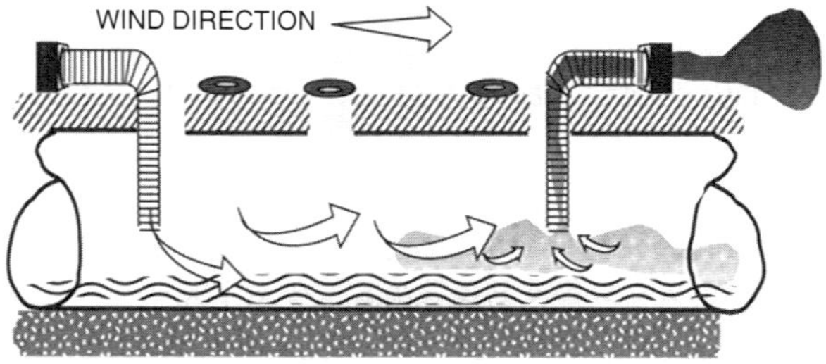

Figure 5-37: Ventilating a confined space with heavier-than-air gases and more than one opening. Auxiliary fan is used to supply fresh air and create a positive bubble.

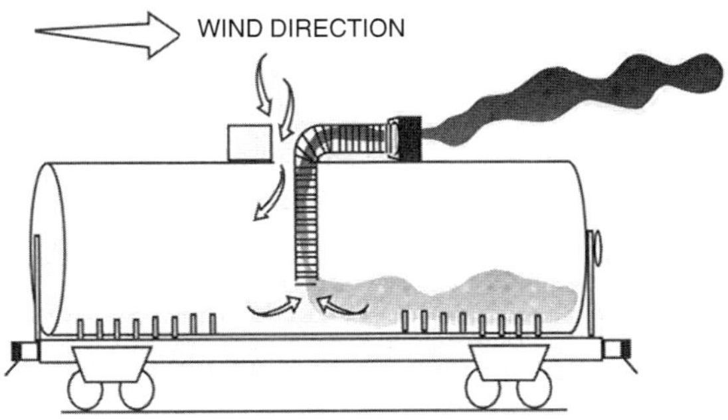

Figure 5-38:
Ventilating a confined space with heavier-than-air gases and only one opening. Liquid should always be completely removed, if possible.

Caution must be exercised when using gasoline powered fans for confined space ventilation. Never use a gas fan that does not have the capability of removing the exhaust from the area of the fan at least 6–8 feet (Figure 5-39). Regular monitoring of the space for carbon monoxide and other by-products of combustion is required.

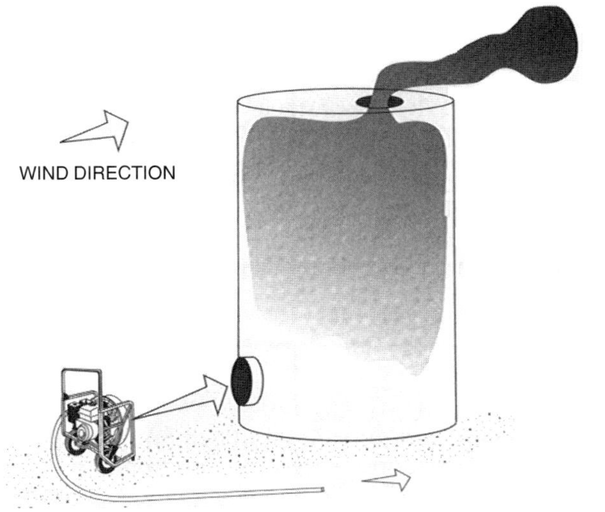

Figure 5-39: Gas powered fan used for confined space ventilation showing remote exhaust.

Use a metal remote exhaust hose extender for gas powered fans used in confined space ventilation.

When only negative pressure ventilation is used for confined spaces, it is possible to pull in contamination from unknown sources, such as from the ground or through small pipes (Figure 5-40). Continuous monitoring of the atmosphere of the space for a number of possible contaminants is required.

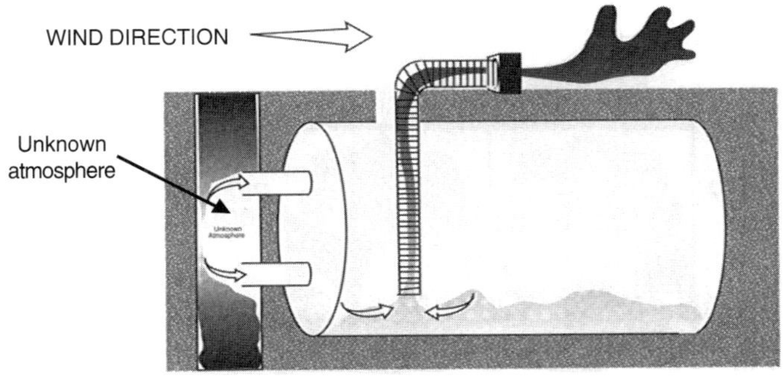

Figure 5-40: Using only negative pressure may pull contamination into a confined space.

High-Rise Buildings

High-rise buildings create special ventilation problems. However, the ventilation principles are the same. Air still moves from an area of higher pressure to an area of lower pressure. Factors included in high-rise ventilation include the following:

FACTORS TO CONSIDER IN HIGH-RISE VENTILATION

- distance from ground level to roof
- number of potential openings
- stairwells and elevator shafts
- occupants
- HVAC systems
- weather
- prefire planning
- manpower

BUILDING HEIGHT AND NUMBER OF OPENINGS

Although the height of a high-rise is a factor in ventilation (it affects the time involved), it does not negate the usefulness of ventilation. Air will always tend to rise in high-rise buildings. This tendency is multiplied in shafts that extend the entire height of the building. Shafts that start and terminate within the building will not flow as much air as shafts that cover the entire building. Because of the large number of doors and windows associated with high-rise buildings, the potential for interruption of air flow and currents created by mechanical ventilation is multiplied accordingly. In some cases, firefighters may be required to police ventilation corridors to ensure that doors and windows are not opened or closed to interrupt air flow.

Air will always tend to rise in high-rise buildings.

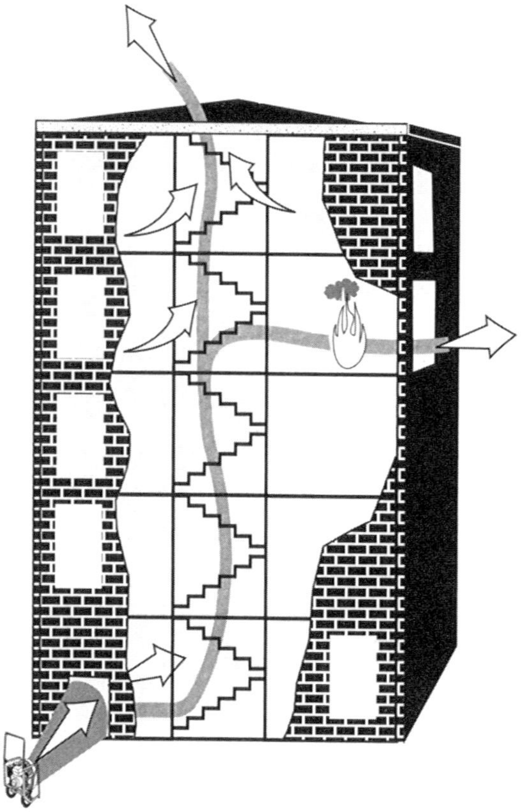

Figure 5-41: Ventilating a stairwell in a high-rise building.

Stairwells are the main arteries of high-rise buildings. They are the means of egress for occupants and the means of ingress for firefighters. When pressurizing stairwells to maintain this egress/ingress route, firefighters must understand that smoke and fire gases are still going to rise to the top floors and mushroom down. If there is not an opening at the top of the stairwell, it may be possible to force smoke through the fire floor to the outside; smoke in the stairwell above the fire will have to be exhausted through the top floor and out (Figure 5-41).

STAIRWELLS

Many buildings have pressurized stairwells built into them. Prefire planning should indicate if stairwells are pressurized and how the HVAC system accomplishes the pressurization. Fire service fans can assist in the pressurization of HVAC pressurized stairwells and/or direct pressurized air.

OCCUPANTS

Because high-rise buildings usually have many occupants, the need for rescue and ventilation operations is increased accordingly. In addition to being a rescue problem, occupants may inhibit ventilation operations by opening doors and windows that fire service personnel have closed to direct air currents.

HVAC SYSTEMS

HVAC systems can be of great benefit for firefighters during the ventilation process. HVAC systems that have not been preplanned, however, can send smoke and/or fire to areas previously unaffected by the fire. In cases where the fire department is very familiar with the HVAC system, fire service fans can augment the system.

WEATHER

Weather is a factor in high-rise buildings, but again it should not stop fire service personnel from performing the ventilation function. Air will rise until it reaches a level where its temperature is the same as the temperature of the surrounding air. Air flow that stops rising at ambient temperature is called the "stack effect." As the air rises, it cools. The greater the distance the air has traveled, the more it cools. The higher the building, the greater the chance that the air will cool to the ambient temperature. Weather affects the ambient temperature; however, tests have shown that in a pressurized situation, air will always tend to rise.

Smoke will rise to the highest available point. If there is no opening to the outside, the smoke will bank or mushroom down to lower levels. If smoke cools to ambient temperature, it will stratify at that level and mushroom down from there.

PREFIRE PLANNING

As already stated, prefire planning is essential in combating high-rise fires. In addition to rooftop openings and HVAC systems, prefire planning should include the types of openings and windows in the building, the time of day that occupants occupy the building, the location of all shafts and hoistways, and the availability of the building engineer in case of HVAC operation.

Through prefire planning, some departments have identified that fire service fans should be strategically located within the building so that they are available during an emergency. They are usually supplied by the building owner and are under the control of the fire department. While gas fans should not be used within buildings, in some cases they may have to be put into service because of the impracticality of advancing power cords to the area needed.

STAFFING

Manpower is a major concern during high-rise firefighting. Additional manpower must be assigned to the ventilation sector to cover the larger amount of area. Additional manpower may also be needed to police ventilation ways so air is not redirected. As with all buildings, smoke removal should begin on the lowest affected floor and progress toward the top.

Communication with the command post should provide continuous status reports, as the ventilation of high-rise buildings is a top priority. Smoke is usually a greater danger than fire and in many cases causes greater problems.

Below-Grade Conditions

Heavier-than-air gases can only be evacuated by mechanical means.

Because heavier-than-air gases by definition do not rise, they create special ventilation problems. These gases tend to accumulate in basements, tunnels, pits, and even drains. They can only be evacuated by mechanical means. Many heavier-than-air gases require ventilation fans with explosion-proof motors.

To evacuate heavier-than-air gases from below-grade spaces, use fans with flexible ducting, and place the inlet to the negative pressure fan at the lowest point possible. Use a positive pressure fan to supply replacement air (Figure 5-42).

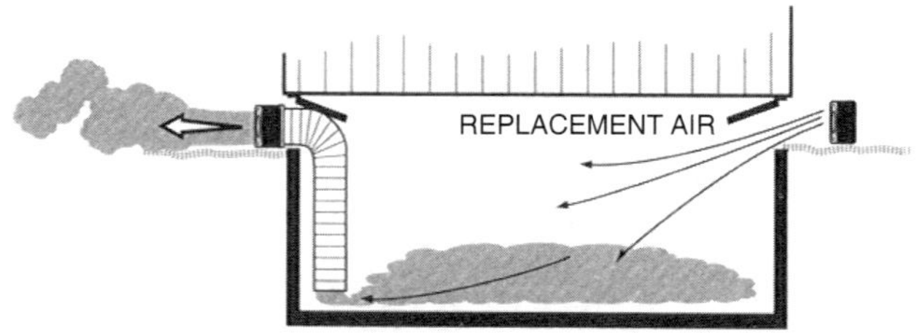

Figure 5-42: Removing heavier-than-air gas below grade.

For lighter-than-air gases in a basement, use ducting to position the replacement air inlet at the lowest point possible, and place the negative pressure fan as high as possible (Figure 5-43).

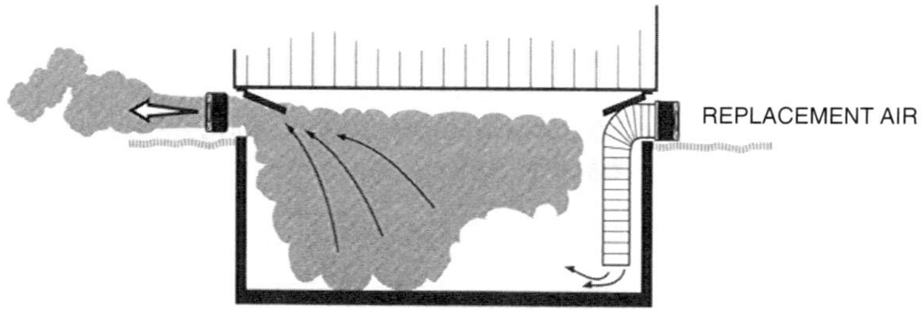

Figure 5-43: Removing lighter-than-air gas or smoke below grade.

For manholes, use two electric fans with flexible ducts (Figure 5-44).

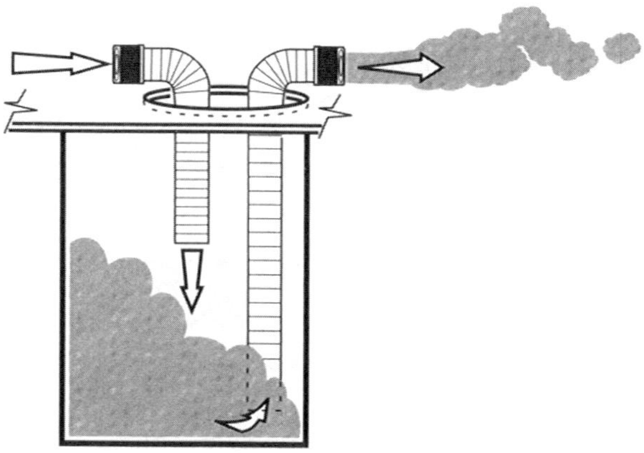

Figure 5-44: Ventilating a manhole.

Before entering below-grade spaces, be sure to ventilate and monitor the atmosphere.

Other Situations

Interior Rooms

Office buildings, computer rooms, and temporary walls set up around a piece of equipment or operation are all examples of interior rooms. When ventilating interior rooms, be sure to allow for replacement air (Figure 5-45). This can be done by providing an opening twice the size of the fan or by using a fan or fan/duct combination to supply air. The pressure in the interior room should be slightly less than the pressure in adjoining rooms. This will prevent smoke or paint fumes from entering adjoining rooms.

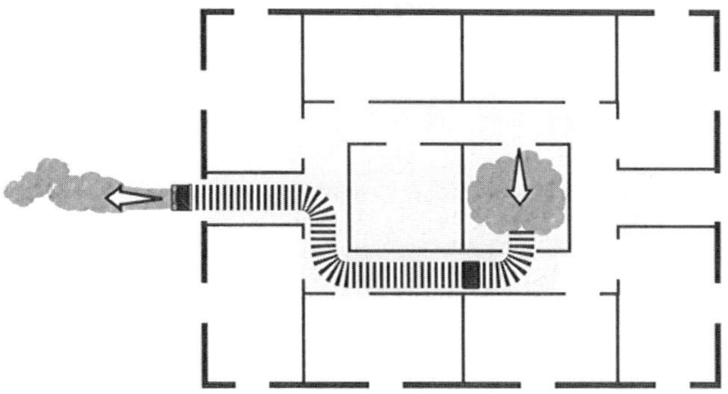

Figure 5-45: Exhausting interior room atmosphere using two electric fans and duct.

Emergencies Involving Temperature

Situations involving hot or cold temperatures range from those that have temperatures causing discomfort to extreme temperatures that can cause death. Electric fans can move warm air to a cold area, or they can move cool air to a hot area during temperature emergencies. Temperature emergencies involve situations such as persons trapped during inclement weather or industrial accidents near or in hot machinery. Fans may also be used in the rehabilitation area to warm or cool firefighters.

Air Movement for Cooling

Water is well known as a cooling or extinguishing agent, but air is more plentiful, is more readily available, and can be extremely effective where properly used. Normally, air with 30% or better relative humidity provides effective insulation against heat and fire. Air in an immediate fire area has been dried out; thus, the fire burns vigorously unless heated air and smoke are removed or displaced. As the temperature increases, the amount of moisture (humidity) near the fire is quickly absorbed. At higher temperatures, the moisture absorption is much greater.

Fog Streams

If cool, moist air (all natural air has some moisture) is introduced into a fire situation, not only is the air itself a coolant, but the natural moisture inhibits combustion. As the temperature is reduced, much of the moisture in the heated air is released and contributes further to suppress combustion. Often this released moisture can be seen in the form of condensation after a fire is extinguished. A quick and efficient way to introduce moisture to dry air and gases in a fire is by using spray or fog nozzles. Water spray cools excessive heat and purges the fire area as water particles expand into steam and replace moisture lost from the heat of the fire.

Moisture in air, whether natural or provided by hose streams, provides a barrier against radiated heat. Observe the protection afforded by a fog nozzle. It transfers heat from the fire area when converted into steam, and it interferes with the oxidation process.

Fog streams have the capability of moving considerable amounts of air; however, they have many drawbacks:

DRAWBACKS OF FOG STREAMS

- Use large quantities of water for non-attack operations.
- Must be continuously manned.
- Often can only be used directly to the outside of a building.
- Have limited range.
- Difficult to get nozzles into effective position in a highly heated area.
- Create additional water damage.
- Require continuous use of a pump.

After flame production ceases, there may still be considerable latent heat, but cool, moist air serves to reduce this heat until fuel distillation ceases completely. While small fog streams stop flame production, they have inadequate cooling capacity to remove the high latent heat remaining in the fuel. Sometimes twice as much water is needed in this "overhauling" stage as in the "flame suppression" stage. Tests measuring the quantities of water needed to control fires with fog nozzles point up this fact.

Exclusive dependence upon water cooling ignores the efficient, practical, and rapid means of getting rid of excessive heat by mechanical ventilation. Mechanical ventilation properly applied quickly removes heat to the outside atmosphere.

To summarize, several basic means of heat reduction are readily available for fire suppression:

MEANS OF HEAT REDUCTION

- Cooling water sprayed to absorb and transfer excessive heat.
- Mechanical ventilation to move heat outside, where it is quickly dissipated.
- Mechanical ventilation to move in cool, moist air to replace heated gases and reduce fuel temperature.
- Cool, moist air to insulate fuel that might be heated by the fire to ignition temperature.
- Cool, moist air to inhibit flame production.
- Cool, moist air to reduce the temperature in the fire area and free latent moisture trapped by excessive heat (remember that the relative humidity in the immediate burning area may approach zero).

The Use of Ventilation Fans With Fog Nozzles

Firefighters combating a blaze with fog equipment typically hold the fog nozzle as high as possible and aim it directly at the source of the blaze. However, in many fire situations, the water particles in fog streams do not have sufficient mass or weight to provide effective range. Also, obstructions such as shelves, partitions, etc., may prevent the fog stream from reaching the fire. These problems can be solved by using a ventilation fan with the fog equipment.

Ventilation fans permit the firefighter to combine the best advantages of both water and air as extinguishing agents. By combining fog equipment with the powerful, spiral air screw produced by the smoke ejector, the natural tendency of water to coalesce is averted. This combination of air pressure and aeration tends to keep the water droplets separate and extends the range and effectiveness of spray streams. Water discharged in this manner can be directed far beyond the reach of a normal nozzle and in a form that can provide more effective extinguishing and purging than by water spray alone.

ADVANTAGES OF COMBINING AIR PRESSURE WITH FOG STREAMS

The following points should be observed for successful firefighting using both ventilation fans and fog nozzles:

- Place the blower on the floor so the spiral blast of air is directed toward the fire.
- Hold the fog nozzle so that it is parallel with the airstream and is just above the edge of the fan. Make sure that the fog stream contacts the airstream 3–4 feet from the exhaust end of the smoke ejector.
- Practice diligently and drill frequently in order to acquire the necessary skill for effective combat of the unique problems present in each activation of fire.

Equipment Use Instructions

Portable Generators

Frequently fire departments must depend on emergency power generation to operate lights, electric drills, saws, and electric smoke ejectors. The following guidelines should be followed to achieve the highest performance when operating electric smoke ejectors on emergency electrical power supplies.

- Smoke ejectors with AC motors are either permanent split phase or capacitor-start motors and are designed to operate at a continuous, constant rpm. It is important that the voltage and cycles remain constant or within plus or minus 3% of specifications. Continued operation outside these limits causes overheating and probable motor failure.
- Selection of the correct generator is important. Not only must it provide adequate amperage to operate all of the smoke ejectors required at one time, but it must supply sufficient amperage to operate during the start-up phase, as the amperage draw is much higher during start-up than during the running phase.
- When running on a generator, smoke ejectors should be plugged in one at a time. When one fan is up to full speed, another may be plugged in.
- Generators must operate at a constant voltage and cycle and must provide the rated amperage.
- Apparatus powered generators should be monitored to provide the proper rpms to produce the required voltage and cycles.

- Use heavy-gauge extension cords. Small-gauge extension cords reduce voltage to fans.
- Keep the distance between fans and generators to a minimum.

Extension Cords

Extension cords work in much the same way as fire hoses. More pressure is required for 100 gpm at the end of a 400-foot 1-1/2" hose than for 100 gpm at the end of a 50-foot 1-1/2" hose. Similarly, the longer the extension cord, the less voltage there is at the workable end. This problem can be overcome by using extension cords with heavier wires which conduct more electricity and result in a lower voltage drop. The voltage drop caused by undersized extension cords usually cuts the performance of the smoke ejector to below the motor nameplate specifications. See Figure 5-46 for a list of recommended extension cords. WARNING: Never exceed cable extension lengths. Always use the required wire gauge. Never use reeled or coiled cable. The power supply must handle wattage requirements.

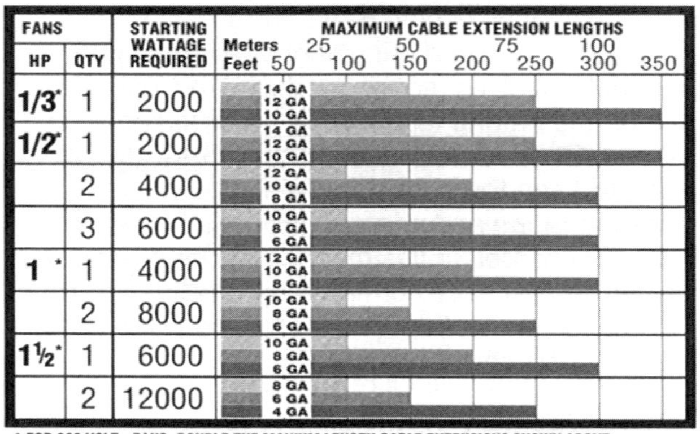

* FOR 220 VOLT - FANS, DOUBLE THE MAXIUM LENGTH CABLE EXTENSIONS SHOWN ABOVE.

Figure 5-46: Recommended extension cords.

Cord Reels

Care must be taken when using cord reels to power any portable electric equipment. If the cord is not completely unrolled, the unrolled part of the cord produces a counter-EMF (electromotive force). The counter-EMF in effect reduces the current-carrying ability of the cord. If the cord size and length are near the limits recommended above, either pull the entire length of cord off the reel when using portable equipment or replace the cord with a larger cord (Figure 5-47).

Figure 5-47: Proper use of a cord reel.

Safety Precautions

The following list of safety items must be read thoroughly by anyone using ventilation equipment.

- This chapter contains an overview of safety standards. Firefighters, chiefs, and confined space entry professionals must understand these precautions and ventilation fundamentals before using this equipment.
- Never enter an area without monitoring the atmosphere and/or wearing self-contained breathing apparatus.
- No fan is explosion proof. The fan motor may be, but any rotating device (blade) that is in close contact with a stationary device (housing) can cause a spark, either by static discharge or through a small metallic item entering the fan guards.

No fan is explosion proof.

- All gasoline powered fans produce carbon monoxide and other emissions. When gasoline powered fans are used, this must be taken into account.
- Never use a fan with a damaged electrical cord or plug.
- In combustible atmospheres, place the smoke ejector in position with the switch in the OFF position. Plug the unit into an outlet that is in a safe environment, as far away as possible from the fan. Turn the switch ON. When the switch is turned on, the starting arc is contained inside the motor. Never operate a unit in a combustible atmosphere if the motor has been tampered with.
- Fans must never be operated or adjusted while the guards are removed from the unit.
- Always unplug the fan before cleaning or removing the guards. Units are provided with an automatic overload that allows the motor to restart after it has cooled down.
- The guard with the smaller spacing must be attached to the blade side of the fan to prevent accidental contact between fingers and the blade.
- Warning labels are key equipment parts. Do not remove, change, or cover these labels. If the labels are not readable, contact the manufacturer for new ones.
- An electric fan must be properly grounded.
- Prevent heavier-than-air gases from entering other low areas where damage or injury could occur. Drive gases forcefully upward into the atmosphere where they can be diluted.
- Prevent paper, plastic, and cloth from covering the intake of a fan.
- Do not operate fans too long in severe cold weather, especially if the fire has affected the heating system. Prolonged operation in such conditions can cause water pipes to freeze.
- Pick up the fan **only** by its handles or hangers, using caution and good lifting practices.
- When a blowing fan is on the floor, supervise traffic in the area. The suction of the fan can draw loose clothing into the unit. Keep children away from any operating fan.

Fan Maintenance

As with any fine piece of equipment, a fan that is properly maintained will perform better and last longer.

- Keep the fan blade clean. If soot and tar are not removed, accumulations will cause unbalanced blades, which can create excessive vibration and motor bearing failure. Following use, remove fan guards and clean the blade, motor, and inside of the housing with detergent and water. If necessary, use solvents. Do **not** use abrasives, sharp tools, knives, screwdrivers, etc. These tools may score the blade, thereby causing imbalance or weakening. Fan blades are made of cast aluminum, so do not use caustics that attack aluminum.

Fan guards are spaced for maximum safety and should always be on the ejector.

- There are fan guards on the intake and discharge sides of the smoke ejector. They are **not** interchangeable. When replacing guards after cleaning, position them with the rings to the outside and with the row of spot welds holding the rings on the bottom. A missing hold-down bolt can allow a corner of the guard to vibrate, permitting a finger to get behind the guard and into the fan. Fan guards are spaced for maximum safety and should always be on the ejector.
- Always keep the rubber feet on the ejector. If they are removed, the thrust developed by the blade causes the unit to move along the floor.
- Proper maintenance of gas powered fans is very important and needs to be done on a routine basis. The oil level must be checked prior to operation. Be sure the octane level of the gasoline is acceptable. On two-cycle engines, also be sure the oil and gas mixture is correct. Sparkplugs should be examined periodically, and carburetor jets must be adjusted for maximum rpm. Cold weather may affect engine operation.

Other Uses for Fire Service Fans

Community Services

- As a public service, use ventilation fans to remove obnoxious odors, fumes, gases, refrigerants, and fumigants in homes, restaurants, offices, businesses, storage areas, etc.
- Use the fans to cool by increasing circulation in these situations:
    - Overheated rooms or buildings not involved in fire, especially those occupied for public assembly purposes.
    - To improve working conditions when firefighters are cleaning up.
    - To cool electric motors when they overheat and when normal ventilation around the motor is inadequate.
    - On generators, transformers, and condensers when needed or requested.
    - On internal combustion motors to remove exhaust gases and to cool engines to prevent overheating.
    - On any machinery which becomes abnormally warm and for which additional circulation will be advantageous.
    - In refrigerated rooms to distribute low, colder air evenly over all products for even chilling or freezing, especially during periods of hot weather or when the plant is overloaded.

- Aid in evaporation:
  - In drying newly finished walls and floors when they won't dry properly.
  - To distribute deodorizing agents to absorb odors.
- Aid law enforcement agencies by removing tear gas, "stink bombs," and other gases rapidly and effectively.
- Supply fresh air to mine shafts, holds of ships, tunnels, caves, and below-grade places.
- Permanent installations:
  - Exhaust systems for restaurants, manufacturing plants, industrial plants, and large storage buildings.
  - Dust-collecting systems for cement in concrete-mixing plants.
  - Exhaust systems in parking garages.
  - Cooling and circulation systems in any building to increase comfort and efficiency of employees.
  - Roof and attic ventilating units for homes to cool them for more comfortable living.

Neutralizing and Deodorizing

An extensive study of smoke odors shows that quick and complete removal of smoke eliminates most smoke odors. The degree of effectiveness is determined by the speed and thoroughness of smoke removal upon arrival of the fire department. When damage is extensive or smoke has cooled and condensed upon the building and its contents, there are large smoke-damage problems. By making smoke removal a part of basic firefighting, fire departments can render an invaluable service.

To be practical for fire departments, deodorants or neutralizers must possess the following qualities:

QUALITIES OF DEODORANTS AND NEUTRALIZERS

- Be non-allergenic to the occupants.
- Be safe (non-toxic) for firefighters and occupants during application and after firefighters have left.
- Be relatively inexpensive and readily available.
- Be a product with which the occupants are familiar and can understand so they can reapply it after the firefighters have gone.
- Be fast and easy to apply so the job is completed in a relatively short time.
- Be safe, reliable, and leave no objectionable residual odors. Unsatisfactory results can lead to liability suits against the fire department (especially since deodorizing is not considered an emergency in many cases).

Average household deodorants work very well. They meet most of the above standards and are safe for use by firefighters. When conditions require a stronger deodorant, the job should be done by professional deodorizers. They have the necessary chemicals, applicators, time, and insurance for complete service.

The following is an outline of the process used by two national deodorizing firms:

DEODORIZING PROCEDURE

1. Remove smoke as quickly and thoroughly as possible.
2. Remove heavily damaged material such as charred furniture, wood, and drapes.
3. Apply deodorant freely on and around all overstuffed furniture, closets containing clothes, drapes, curtains, and on charred wood, or as directed on the container of the deodorant.
4. Use ventilation fans to circulate and distribute the deodorant throughout the area with all openings closed.
5. After deodorants have taken maximum effect, place ventilation fans in door and window openings to expel deodorant agent from the area. Open other doors and windows to obtain a cross-draft.
6. Repeat procedure as necessary.

Salvage and Overhaul

The protection of property during and after a fire is referred to as "salvage." Overhaul activities are designed to ensure that all elements of fire have been fully extinguished and that there is no potential for any further ignition or rekindling of fire from the original event. Salvage reduces the total property loss in the building by protecting those items that may be damaged during the firefighting operation. Typical salvage tasks include covering furniture to prevent smoke and water exposure, protecting fragile and valuable items, rerouting water, and relocating photos, jewelry, and vital records. Overhaul includes searching for hidden fire and ensuring that concealed spaces have been thoroughly checked. These activities can garner great public support and communicate the attention to detail that your personnel took to make a difference on the occupants' worst day, ever.

Summary

Offensive fire attack is the strategy in which fire companies coordinate their activities to initiate fire attack, conduct searches, ventilate, perform needed rescues, salvage undamaged property, and ensure against further rekindle. The first-in company officer must be ready to accept any of these broad assignments and manage the assigned resources to effect a positive result.

Note: The Command Worksheet on the next page is for use with the student exercises that follow. This sheet may be duplicated as needed.

COMMAND WORKSHEET

INCIDENT INFO

Incident Name ______________________

Time/Alarm	Address	Unit No.	Inc. #	Date
Time/Scene	Occupancy – Size/Type/Acres		Property Loss	
1st. Scene	Occupant	Telephone	Contents Loss	
Control	Owner	Telephone	Civ. Inj.	F.D. Inj.

RESOURCE STATUS

Company	A/S	Assignment

☐ INITIAL INCIDENT ACTION PLAN

1 ☐ Offensive ☐ Marginal ☐ Defensive

2

LIFE SAFETY OBJECTIVES	INCIDENT CONTROL OBJECTIVES	PROPERTY OBJECTIVES
☐ ☐	☐ ☐	☐ ☐
☐ ☐	☐ ☐	☐ ☐
☐ ☐	☐ ☐	☐ ☐

3 # Co's Req'd ___ # Co's Req'd ___ # Co's Req'd ___

Plus R.I.C. ___ → TOTAL # CO'S ___ # ALARMS ___

4 Communications Plan ☐ Command ☐ Tactical ☐ Air to Ground Tac

5 Level II Staging Area Location

Rescue		Vent	
Yes	Clear	Smoke	Clear

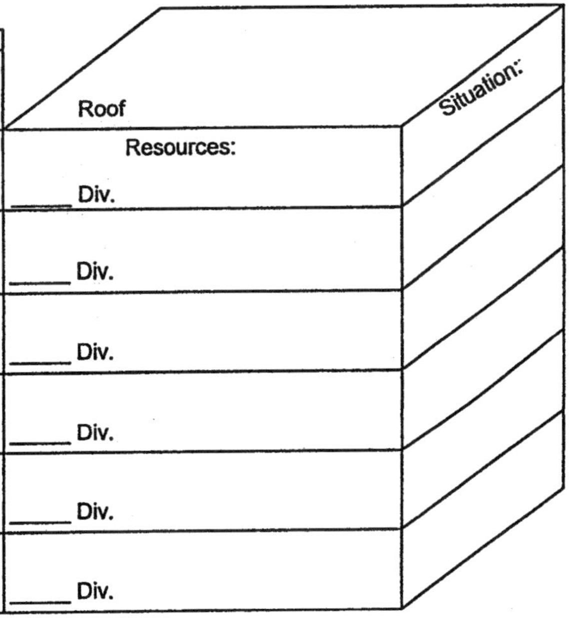

TACTICAL BENCHMARKS

- ☐ Fire attack
- ☐ Water supply
- ☐ Primary search
- ☐ All clear
- ☐ Utilities
- ☐ Ventilation
- ☐ Back-up lines
- ☐ Ext check / exposures
- ☐ Salvage
- ☐ Secondary search
- ☐ All clear
- ☐ Progress reports
- ☐ Rescue laddering
- ☐ Sprinkler support
- ☐ Breathing / lighting support
- ☐ Rehab
- ☐ Traffic control
- ☐ P.I.O. / media notification
- ☐ Under control
- ☐ Investigator
- ☐ Utility co's
- ☐ Building rep
- ☐ Building and safety
- ☐ Release plan

Student Exercise 5-1

In this exercise, the student is playing the role of the company officer on the first-arriving apparatus. The student should complete the following:

1. Receive the initial dispatch information.
2. View the written and graphical prefire plan information.
3. Conduct a size-up from the initial view upon arrival.
4. Transmit a radio size-up.
5. Develop an ICS chart for the initial incident organization.
6. Assign all first-alarm companies by radio.

Dispatch Information

You have just been dispatched to a residential subdivision. From the address, you realize that this neighborhood was constructed within the past six years. The residence is wood frame, with lightweight trusses and a tile roof. The time of alarm is 0359 hours. The temperature is 68°F, and the humidity is 18%. Winds are from the south at 5 mph.

"Battalion 1 ... structure fire. Attention, Battalion 1 units and stations. A residential structure fire, 2462 West Second Street. 2462 West Second Street. Cross street: Willow. Respond on channel 2 at 0400 hours."

Arrival Information

Upon arrival, you find fire and smoke showing from a three-car attached garage. Fire has not vented the roof.

You are responding with your department's first-alarm assignment. The hydrants are spaced according to your department's residential subdivision ordinance. The instructor will advise on hydrant location.

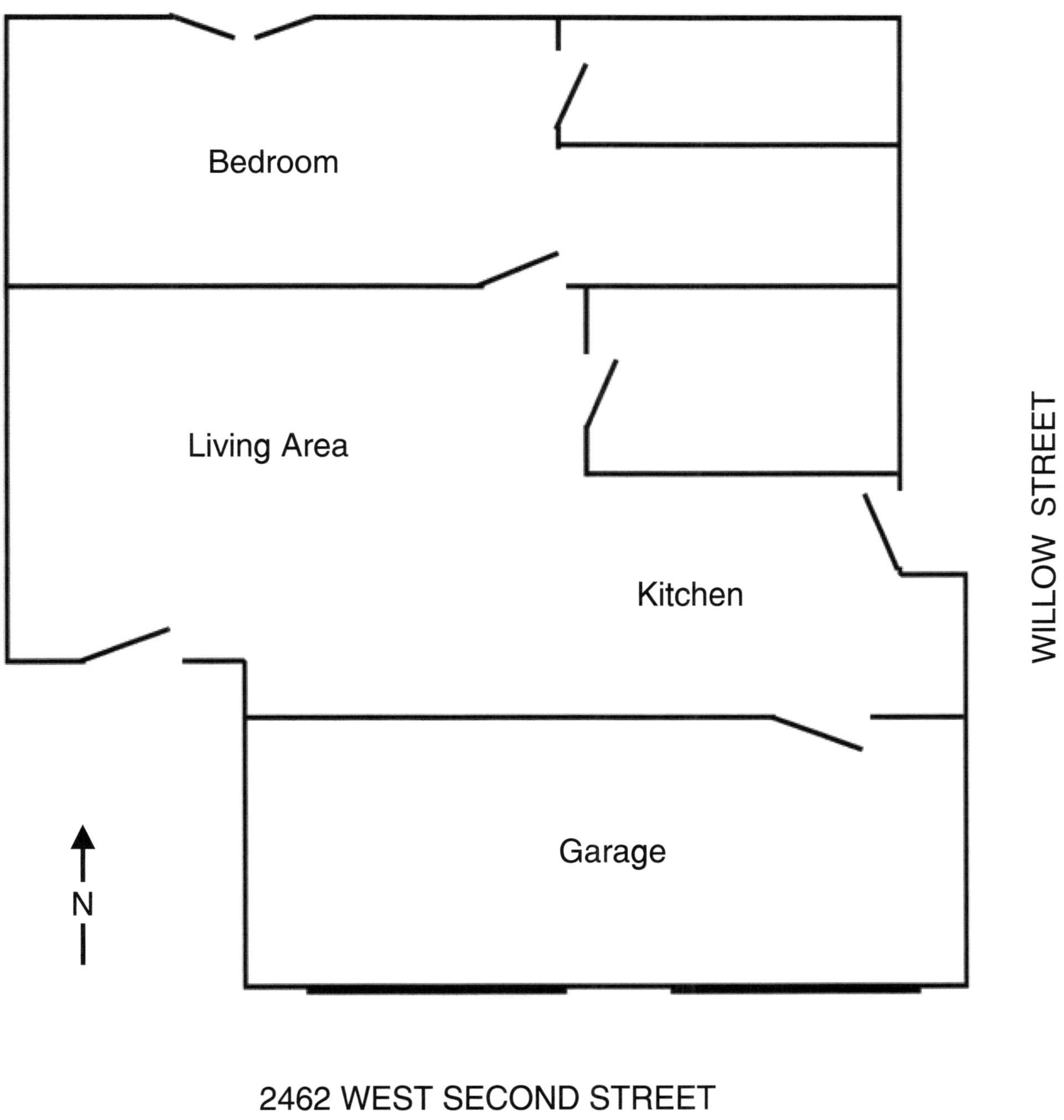
Bedroom
Living Area
Kitchen
Garage
N
WILLOW STREET
2462 WEST SECOND STREET

Student Exercise 5-2

In this exercise, the student is playing the role of the company officer on the first-arriving apparatus. The student should complete the following:

1. Receive the initial dispatch information.
2. View the written and graphical prefire plan information.
3. Conduct a size-up from the initial view upon arrival.
4. Transmit a radio size-up.
5. Develop an ICS chart for the initial incident organization.
6. Assign all first-alarm companies by radio.

Dispatch Information

You have just been dispatched to a residential subdivision. From the address, you realize that this neighborhood was constructed within the past six years. The residence is a two-story, wood frame building, with lightweight trusses and a tile roof. The time of alarm is 0235 hours. The temperature is 68°F, and the humidity is 18%. Winds are from the south at 5 mph.

"Battalion 3 … structure fire. Attention, Battalion 3 units and stations. A residential structure fire, 184 Bryant Road. 184 Bryant Road. Cross street: Megan Court. Respond on channel 3 at 0238 hours."

Arrival Information

Upon arrival, you find fire and smoke showing from an upstairs bedroom window. Fire has not vented the roof.

You are responding with your department's first-alarm assignment. The hydrants are spaced according to your department's residential subdivision ordinance. The instructor will advise on hydrant location.

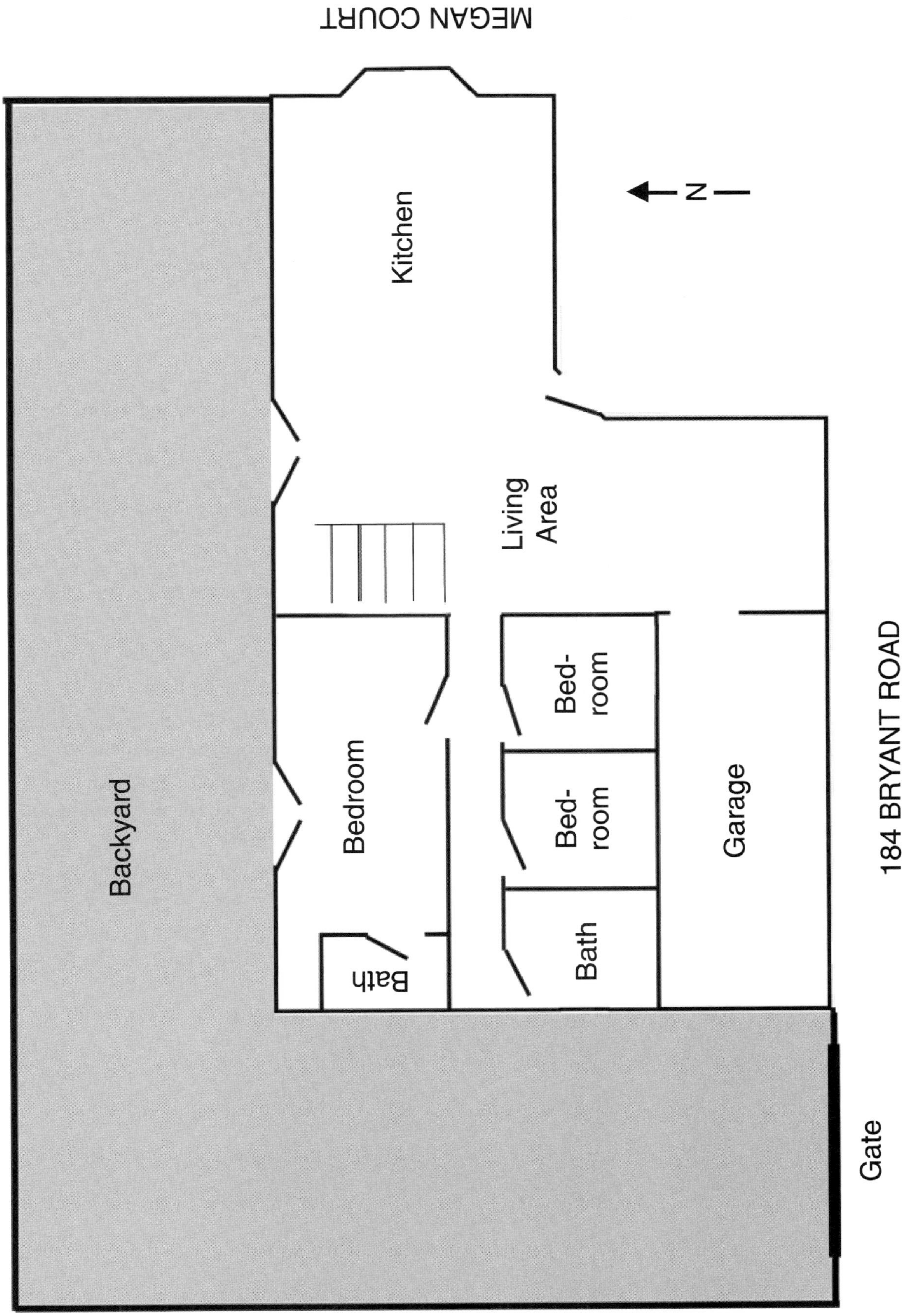
MEGAN COURT
N
Kitchen
Living Area
Bedroom
Bath
Bed-room
Bed-room
Bath
Garage
184 BRYANT ROAD
Backyard
Gate

Student Exercise 5-3

In this exercise, the student is playing the role of the company officer on the first-arriving apparatus. The student should complete the following:

1. Receive the initial dispatch information.
2. View the written and graphical prefire plan information.
3. Conduct a size-up from the initial view upon arrival.
4. Transmit a radio size-up.
5. Develop an ICS chart for the initial incident organization.
6. Assign all first-alarm companies by radio.

Dispatch Information

You have just been dispatched to a mixed residential subdivision. From the address, you realize that this neighborhood was constructed within the past 10 years. The occupancy is a multi-family residence for senior citizens. All residents are ambulatory. The building is a three-story, center-hallway construction. It is constructed of a wood frame, with lightweight trusses and an asphalt/composition roof. The time of alarm is 0554 hours. The temperature is 45°F, and the humidity is 22%. Winds are from the southwest at 10 mph.

"Battalion 1 … structure fire. Attention, Battalion 1 units and stations. A multi-family residential structure fire, 1806 West Main Street. 1806 West Main Street. Cross street: Central Avenue. Respond on channel 2 at 0556 hours."

Arrival Information

Upon arrival, you find fire and smoke showing from the third floor balcony. Fire is auto-exposing an upper floor. The fire has not vented the roof.

You are responding with your department's first-alarm assignment. The hydrants are spaced according to your department's residential subdivision ordinance. The instructor will advise on hydrant location.

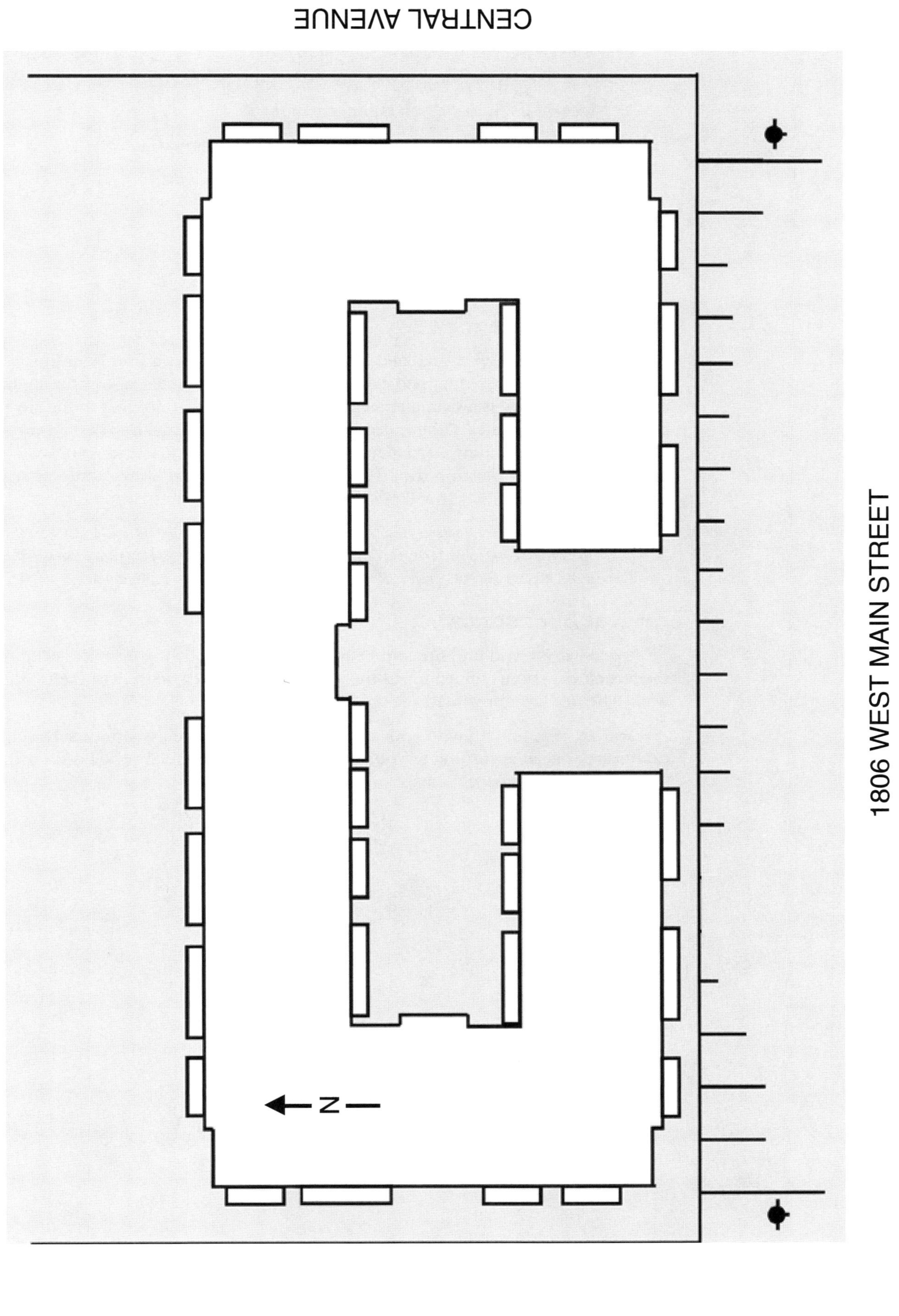
CENTRAL AVENUE
1806 WEST MAIN STREET
N

Student Exercise 5-4

In this exercise, the student is playing the role of the company officer on the first-arriving apparatus. The student should complete the following:

1. Receive the initial dispatch information.
2. View the written and graphical prefire plan information.
3. Conduct a size-up from the initial view upon arrival.
4. Transmit a radio size-up.
5. Develop an ICS chart for the initial incident organization.
6. Assign all first-alarm companies by radio.

Dispatch Information

You have just been dispatched to a commercial structure fire. From the address, you realize that this medical complex was constructed about five years ago. The occupancy is a women's health care clinic. The building is a one-story, commercial strip center. Other occupancies are an ice cream shop/restaurant at the west end. The building is wood frame, with lightweight trusses, exterior stucco, and an asphalt/composition roof. The time of alarm is 1434 hours, temperature is 65°F, and humidity is 41%. Winds are from the northeast at 2 mph.

"Battalion 2 … structure fire. Attention, Battalion 2 units and stations. A commercial structure fire, 62 Medical Center Parkway. 62 Medical Center Parkway. Cross street: Traci. Respond on channel 3 at 1435 hours."

Arrival Information

Upon arrival, you find fire and heavy black smoke showing from the front of the structure. There is burning all the way to the ground, on the concrete sidewalk. The fire has not vented the roof.

You are responding with your department's first-alarm assignment. The hydrants are spaced according to your department's commercial subdivision ordinance. The instructor will advise on hydrant location.

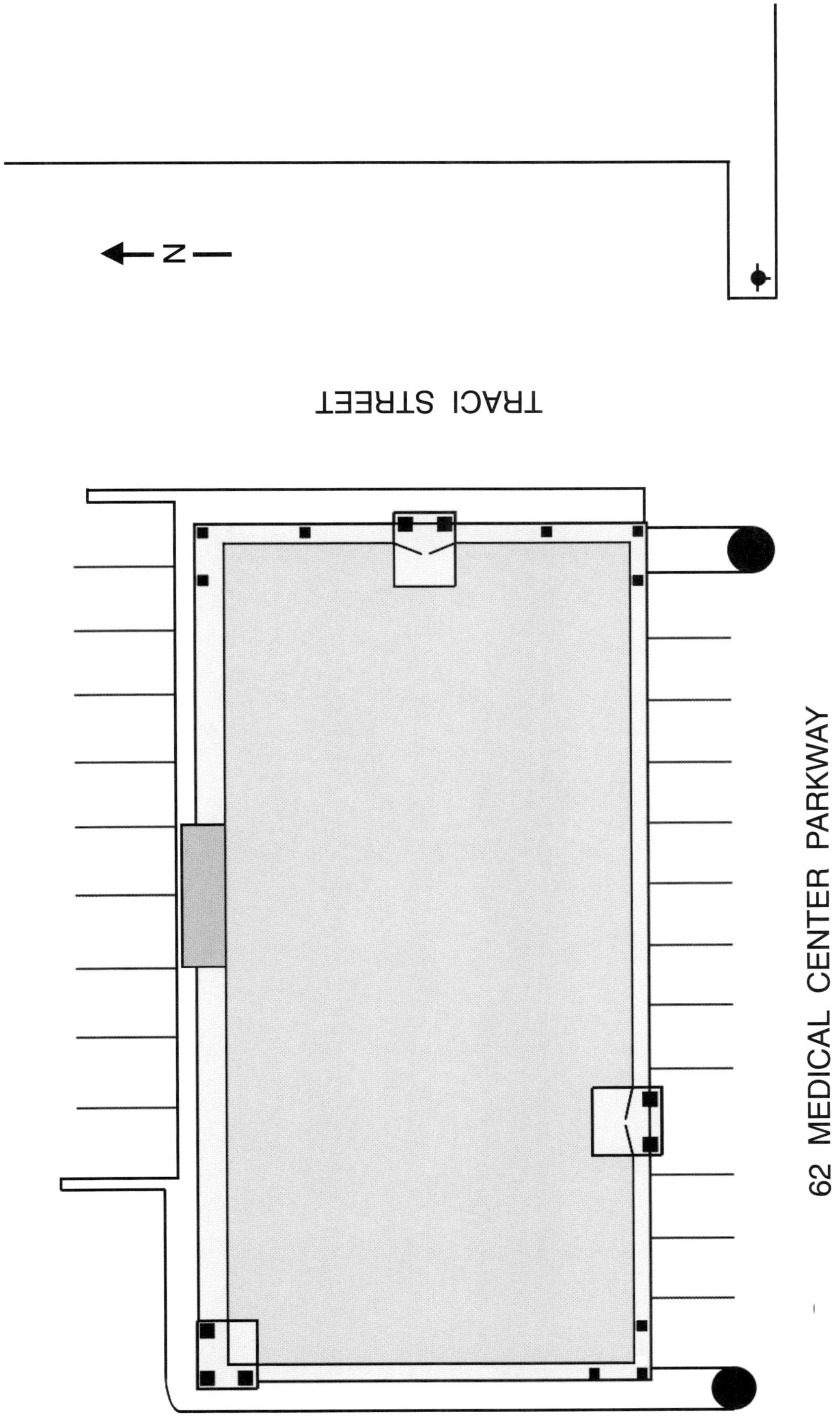
N
TRACI STREET
62 MEDICAL CENTER PARKWAY

MODULE

6

Standards of Fire Attack: Defensive Operations

Objectives

After completing the reading and activities in this chapter, students should be able to:

OBJECTIVES

- Identify and understand the roles and responsibilities of the company officer when assigned as a division or group supervisor on all-risk incidents.
- Implement tactical operations related to exterior fire flow objectives.
- Implement tactical operations related to exterior exposure protection objectives.

Initiating Defensive Fire Attack

One of the most difficult decisions that a fire officer must make is the decision to initiate a defensive, exterior attack on a well-involved structure fire. If an interior attack is chosen instead, firefighters may find themselves in an offensive position under defensive conditions. Trying to maintain such a position is one of the most common contributing factors to firefighters being killed on the fireground.

In EMS situations, we have less trouble recognizing the reality that the traumatic event or the disease process has resulted in death and that there is little intervention we can give to reverse the patient's outcome. On a hazardous materials call, where unknown materials have spilled, we have the discipline to conduct a risk/benefit analysis, and we willingly hold in a defensive mode until the identification is complete. Likewise, in a confined space rescue situation, or a trench rescue scene where multiple victims are trapped in an unstable environment, we have learned that the risk to our own personnel is too great to effect an impulsive rescue attempt. Experience has proven that the best course of action is to stabilize the environment and the atmosphere and to wait to enter the space until all of our backup teams and specialists have determined that it is safe to enter.

In contrast, our fire service culture, our competitive personalities, and our confidence in our own capabilities and equipment lead us to believe that we can beat just about any fire situation.

Far too often, we underestimate the fire's intensity and its effect on the structure, and we overestimate our proficiency in setting up and initiating a large-scale fire attack. Even when we have a good water supply, a quick engine setup, and aggressive firefighters on the line, the structural members often collapse and floors and ceilings fail, causing entrapment. In addition, flashover often creates an untenable interior condition.

The fact remains that the risk to firefighting personnel is not worth the benefit of saving a structure with a large volume of fire in it. Another factor that contributes to firefighter fatalities is that every time we "dodge the bullet" and successfully survive a defensive fire in which an interior offensive attack was wrongfully conducted, we reinforce in our own minds and in our own experience bank that we followed the right course of action. This repetitive reinforcement is one of the things that must change if we are to improve the record of firefighter fatalities.

Every fire company officer and every command officer must refine the skills of identifying defensive conditions. More importantly, officers must direct a defensive, exterior fire attack when the conditions merit it.

Impact of the Community Fire Protection System

The community—not the fire department—sets the level of fire protection for the community.

The "health" or condition of the community fire protection system also plays a role in how fast or how early a fire officer decides to go defensive. Every fire officer must understand that the community—not the fire department—sets the level of fire protection for the community. Through the budget and election process, the community decides whether fire protection will be priority or not. The allocation of resources for fire department staffing, equipment, training, maintenance, and fire prevention will affect the conditions that engine and truck companies face upon arrival. Also, this community decision will dictate the level of involvement that each department can effectively and safely confront. For some, a single-family dwelling of 1,500 square feet that is 50% involved will be the largest fire that can be attacked offensively with the resources provided. For another fire department, the limit may be several floors of a high-rise building, well-involved.

There is a direct correlation between the lack of fire protection resources and the community fire loss.

When the community decides through its elected and appointed officials that it is comfortable with a poor water supply system, no prefire planning program, limited fire department staffing, obsolete equipment, inadequate building codes, and a lack of fire prevention enforcement activities, it is **not** the responsibility of the firefighters and fire suppression officers to risk their lives *unduly* to protect property. There is a direct correlation between the lack of fire protection resources and the community fire loss. If the system stakeholders won't provide a safe level of fire protection, there is no reason for firefighters to die trying to balance the lack of support. If you work in a system that has many of the above components, it would be prudent to take a defensive approach even sooner to protect the lives of your firefighters.

Recognizing Defensive Conditions

Defensive conditions are not always easy to recognize. It is difficult to identify conditions that are "universally" defensive, but the list that follows gives some of the more common defensive conditions. If there is significant fire showing (defined as greater than 50% involvement) with any of the following present, a defensive attack should be initiated:

CONDITIONS INDICATING NEED FOR DEFENSIVE ATTACK

- delayed alarm
- structural elements involved on arrival
- fire through the roof on arrival
- vacant/condemned structure
- no *confirmed* occupants
- no sustainable water supply
- no prefire plan
- lightweight trussed construction
- heavyweight trussed construction
- pre-1933 unreinforced masonry (URM) construction
- hazardous materials or flammable liquids present
- inadequate staffing

Concept of Defensive Operations

The concept of operations for a defensive fire attack has been misunderstood for a long time. If we view the uncontrolled fire as the enemy, we can use a military analogy for our plan of attack. In a combat situation, when we have a large, well-trained enemy, it is time to bring out the big guns. In other words:

Big enemy ⇒ Use artillery

Artillery gives us some real advantages. We can use it against an enemy that we cannot safely approach. Artillery has a much greater reach than small arms fire, so we do not have to put our troops in an unsecured area. In addition, artillery delivers more killing power—more bang, from a greater distance, with more safety. On the fireground, using master stream devices and large-diameter supply lines gives us the same effect. A single master stream monitor can deliver the equivalent of three 2-1/2" hoselines. One person can deliver the same flow that it would take 6-8 firefighters to deliver. Like the artillery, the master stream can perform this function from outside the collapse zone, out of reach of danger.

CHARACTERISTICS OF DEFENSIVE OPERATIONS

The concept of defensive operations has sometimes been described as "surround and drown." The idea is certainly not that simple, but this description does encompass some of the main characteristics. First of all, a defensive fire attack is always conducted from the exterior. This should ensure that all the attackers are outside the designated collapse zone. It is essential to protect exposed buildings from the involved buildings. In a defensive attack, large volumes of water are used from the best vantage points. The plan is to drown the fire, not the neighborhood. Lastly, elevated streams work best after the roof has failed.

Size-Up

The size-up process for a defensive attack is no different from that used for any other incident. There must be a continuous mental evaluation as the fire officer assumes command, transmits the defensive strategy to incoming companies, establishes a remote staging area, and requests additional resources.

Fire Flow Formulas

As we discussed in previous chapters, the NFA fire flow formula can take the guesswork out of initial fire attack flows. It can also be used as a basis for requesting additional resources. Most fire flow formulas assume that one firefighter is required for every 100-150 gpm required.

Resources

What kind of resources are required for a sustained, defensive fire attack? Obviously, engines and trucks are needed to deliver the water through master streams and elevated streams. Other resources should include the following personnel:

- Safety Officers
- Public Information Officers
- Rehab personnel
- Water Supply Officer
- Staging Area Manager
- Division/Group/Sector Supervisors
- Incident Communications Unit
- Rapid Intervention Team
- EMS personnel

Timeline

The focus of a defensive operation is on safeguarding firefighters and protecting exposed property.

The timeline for a defensive operation is dramatically different from that used in most offensive fire attacks. The assumption is that the defensive fire is a sustained, methodical operation. An "all clear" or a "no all clear" signal has already been transmitted, meaning that both the primary and the secondary search are complete. Any actual or potential victims have been rescued or are assumed not to have survived the fire conditions. No search activities will be initiated. The focus of the operation will be on safeguarding firefighters and protecting exposed property. The operation will not be rapid or aggressive. It will instead be safe and efficient.

Marginal Fire Attack

The marginal fire attack is a transitional strategic operation. It is a strategy that is an initial holding pattern until some milestone has been achieved. Marginal fire attack is either an offensive/defensive or a defensive/offensive attack. After the objective or milestone is reached, the operation transitions out of the holding pattern into the preferred strategy. Most fire officers will opt for going offensive if the situation is not clearly defensive. The marginal fire attack is a preferred alternative to going offensive too early with an underpowered attack, only to have the first-arriving chief officer pull the crews out. It is difficult, but company officers must learn to resist the temptation to go offensive when a marginal fire attack is called for.

EXAMPLE OF OFFENSIVE/DEFENSIVE MARGINAL ATTACK

For example, suppose a single, four-person engine company arrives on the scene of a residential fire. There is 25%–30% involvement and a confirmed occupant trapped. The officer decides on a marginal fire attack with an offensive/defensive approach. He transmits a full first-alarm assignment and calls for additional resources. He orders a fire attack with the deck gun and a portable interior

Blitzfire® monitor. The objective is to hold the fire back from the victim's location. The officer then accompanies one firefighter in for search and rescue to remove the trapped victim. After the victim is removed, the officer transitions to a defensive operation and protects exposures. No further interior or offensive operations are conducted, and all subsequently arriving personnel respect the collapse zone and operate in defensive mode.

EXAMPLE OF DEFENSIVE/OFFENSIVE MARGINAL ATTACK

Here's another example: An engine and truck company arrive at an unoccupied strip commercial occupancy that is 30% involved. The building is constructed of lightweight trusses. The total staffing is six personnel. The first-arriving officer decides on a marginal fire attack: defensive/offensive. She transmits a second alarm. Then she initiates a deck gun attack and establishes a large-diameter supply line. The truck company ladders the uninvolved portion of the structure and performs utility control and forcible entry on the adjacent, exposed occupancies. No personnel are allowed to enter, and no offensive operations (interior fire attack, search, ventilation, and salvage) are permitted until the subsequent resources arrive to provide rapid intervention, sufficient fire flow, backup lines, and an extra roof team. As companies arrive, the commander transitions to an offensive fire attack and coordinates a strip ventilation operation with multiple interior attack teams.

In order to choose a marginal fire attack properly, the fire officer must have a realistic expectation of fire company and department capabilities and must have a greater understanding of the fire's potential. The marginal strategy is not an indecisive mode. It is a planned operation to maximize effectiveness despite a limitation in resources. Fine tuning your size-up skills is essential to be effective in using this technique.

WHEN TO USE MARGINAL FIRE ATTACK

The marginal fire attack is a holding pattern for rescues in progress, searches, and the arrival of additional resources. Fire officers should confidently make this decision and should avoid the temptation of going offensive! The marginal fire attack is not an aggressive attack. It is planned and deliberate. A minimum commitment of personnel should be deployed. The strategy requires an enhanced focus on firefighter accountability, as many firefighters will be pulled into a more offensive posture while waiting for the defensive activities to be assigned.

Transition of Strategy: The Operational Retreat

One of the most dangerous situations on the fireground is the operational retreat. This occurs when an offensive fire attack proves to be unsuccessful and the incident commander chooses to stop all interior operations and move to an exterior, defensive operation.

DANGERS ASSOCIATED WITH AN OPERATIONAL RETREAT

Firefighters frequently get lost or trapped during this process. One reason for this is that communication with interior crews is difficult. Some companies and personnel are not willing to give up the ground they have fought so hard to win. Overconfidence and tunnel vision cause some interior officers, who are not familiar with the "big picture," to continue their forward progress despite the order to retreat. They often think, "Just a few more minutes, and we'll have this thing under control!" A level of discipline must exist throughout the organization, so that when the order to retreat or pull out is given, 100% of the members comply—immediately!

Secure Safe Portals for Egress

When the order to retreat is given, operational commanders must ensure that safe portals of egress are created. Moving fire streams to protect these "exit paths" is a prime priority. Fire control must continue until all personnel are moved from the forward area back out to the new attack zone. A collapse zone of at least 1-1/2 times the height of the building must be established and maintained throughout the defensive operation. Be aware that personnel will be more prone to reenter the newly established collapse zone if they have previously been working in it. Enforcement of the zone and maintaining fire company accountability are critical to promoting firefighter accountability. Deputy Chief Jim Murtaugh of the New York Fire Department refers to this phenomenon as "firefighter creep." If allowed to do so, firefighters will enter the collapse zone by a few feet with a hoseline. Then, if not challenged, they will move a little closer, and in a few minutes, a little closer. In a short period of time, they will be right under the eaves and directing streams through a window, oblivious to the collapse potential and the dangers which may confront them.

ENFORCING THE COLLAPSE ZONE

As intact fire companies retreat from interior operations, a roll call or personnel accountability report (PAR) must be accomplished. All personnel should report to a transmitted "rally point" to be accounted for. On a large incident, rally points may be provided on several sides of the incident.

Communications

Radio discipline must be taught, practiced, and enforced.

All forms of communication—face-to-face, runners, and radio—should be integrated to maximize the flow of information to the proper people. Many firefighter deaths have occurred when incident commanders are inundated with useless information during transition periods of the operation and miss critical emergency traffic. Radio discipline must be taught, practiced, and enforced. A designated officer or team should be monitoring the "mayday" channel to ensure that crews that need assistance will immediately receive it.

Sectoring

As the retreat unfolds, a new sectoring of the incident will become necessary. The interior divisions or groups should be terminated. New assignments which provide adequate eyes, ears (intel), and span of control for the commander should be implemented. Continuous evaluation of the command structure must be taken to ensure that accountability, communication, span of control, and incident control objectives are all being met.

Command and Control

MANAGING A RETREAT

Strong and visible command and control must be maintained through the retreat. A calm, confident voice directing companies will prove to be a valuable asset. The retreat should rarely be a surprise or emergency operation. It should instead be planned, anticipated, and actively managed. Company officers can make or break the operation by selective obedience, free-lancing, negative comments, and criticism of the command staff to their subordinate personnel. The time to critique the operation is away from the hazards and emotion of the fire attack. Immediately upon receiving the order to retreat, "blind obedience" is preferred over "Monday morning quarterbacking," when the common goal is the health and safety of all members.

Tools

As in any battle, the weapons and tools available to defeat the enemy are key to a successful outcome. Tools that move large volumes of water, reduce friction loss, have a safety backup, and are flexibly deployed are the tools of choice. Let's look at some of the innovative tools that exist to streamline operations.

Water Supply

The old adage, "When in doubt, lay two out," used to refer to laying supply line. With the advent of large-diameter hose, the wider the better. Knowledge of your community water system is important. Consider the following questions:

CONSIDERATIONS REGARDING THE WATER SUPPLY

- How do the grids work?
- Where are the dead-end mains?
- Are you robbing another engine's flow by taking this hydrant?
- Is there a close hydrant from another grid that is a better choice?
- What is the procedure for boosting pressure from the pumping station?
- Can you supplement the available water with drafting, portable tanks, water tenders/tankers, and other alternative sources?
- Do all pump operators understand the principles of relay pumping?
- Can your department safely pump LDH?

Developing competencies in these basic engineering skills should be added to the company training schedule. Assigning a water supply officer to manage this task can take a large burden off the incident commander.

Appliances

TYPES OF APPLIANCES

Gated wyes, siameses, gated trimeses, stortz adapters, and other appliances don't get used very often, but they may be critical to complete the connection of ladder pipes and later arriving engines and pumps. Knowing what is needed, keeping track of what is available, and ensuring that all seals, O-rings, and valves are in working order should be the responsibility of the engineer or driver/operator, before the fire!

Nozzles

TYPES OF NOZZLES

Nozzles come in a variety of configurations. Selectable-gallonage nozzles give pinpoint control to the firefighter making the attack. Dual-pressure nozzles provide an extra margin of safety during hazardous fire attack situations. Breakaway nozzles make it possible to extend a hoseline without shutting the stream off at the pump. Automatic nozzles have evolved to be lightweight and versatile and to give a perfect pattern on solid or fog applications.

Handlines

HOSELINES FOR DEFENSIVE OPERATIONS

Larger handlines are normally called for in defensive operations. Mobility is not required on most exterior fire attacks. Lines of 2-1/2" or 3" should be handled with a minimum of three personnel. The exception, of course, is the stationary use of an exposure loop. As a rule, 2-1/2" attack lines are the most effective hoselines to be deployed during a defensive operation. There may be a need for several mobile 1-3/4" lines to protect exposures and extinguish spot fires. Volume and reach are essential for a successful defense.

Master Streams

FIXED AND PORTABLE MASTER STREAMS

Master streams can be categorized in two broad groups: fixed and portable. The fixed master stream is permanently mounted to an engine, quint, or fire boat. The position of the apparatus should be out of the collapse zone but close enough to the target to be effective. Portable master streams provide the greatest flexibility in positioning and gaining access in tight spaces such as alleyways or side yards. Master streams are rated to deliver flows greater than 300 gallons per minute. The TFT Saf-Tak® and Crossfire® are proven monitors. Although primarily an offensive weapon, the preconnected Blitzfire® attack monitor can be used as a defensive tool. With flows up to 350 gpm, it can be placed by one or two members and left in place. With the optional automatic oscillation feature (OSC), it is the perfect unmanned monitor which can protect an exposed building or pressurized tank.

Elevated Streams

PREPLUMBED AND DETACHABLE ELEVATED STREAMS

Elevated streams are master stream devices that are attached to fire department aerial ladders and platforms. There are two categories of elevated streams in use today. The first is the preplumbed or permanently attached. The second is the detachable ladder pipe, which must be secured to the fly section of the aerial and a supply line stretched to the tip. Most of the modern, preplumbed appliances can be operated from both the tip and the turntable. The older ladder pipes must be manually operated by a firefighter aloft. Weather and the type of contents involved may limit this application.

CAUTIONS REGARDING ELEVATED STREAMS

Elevated stream operations are dangerous. The large flows, the water load transferred to the compromised building, and the fatigue of personnel late in the incident can all combine to create an increased potential for injuries. Coordination and organization are key. Having an incident command organization that is responsive to the conditions confronting the troops is essential. The incident organization needs to be large enough to plan and execute tactics ahead of the fire. It must also be small enough that information flow throughout the incident is effective.

Coordination of Division/ Group/Sector Officers

During defensive operations, there is a critical need to enhance the coordination between adjacent sectors. Large master streams don't always stay on their side of the sector or division boundary. Opposing hose streams, robbing water and other resources, and deciding when crews get rehab or release all require coordination between sectors.

MANAGING INFORMATION FLOW

The Planning Section is critical for a smooth operation. When polled, most firefighters state that they have never been on an incident where the Planning Section was activated. Most incident commanders keep track of resources themselves (either on the board or via a command worksheet), or they allow Fire Communications to handle the information. In addition, incident commanders generally keep the "situation status" activities (what the fire is doing) to themselves. They walk around and occasionally ask interior crews, "What have you got?" and they feel that they have the planning process under control. On large-scale incidents, having another chief officer take the responsibility for managing the information flow regarding the resources and the situation creates an environment in which only validated, confirmed information gets to the commander. All the "what ifs" are evaluated, and several Plan B and Plan C scenarios are created as contingencies. Planning also coordinates the request for and the release of large amounts of resources and allows the incident commander to focus on strategy rather than on tasks.

Other Defensive Concerns

COVERAGE

On a large defensive operation, coverage must be considered. As fire and EMS resources are committed to a sustained operation, the IC must develop a plan to fill in vacant fire stations. These situations are best developed in an SOP or mutual aid agreement and worked out before the incident.

RELIEF

Making provisions for relief is another time-consuming activity. Relieving a shift at the scene creates all sorts of confusion. Developing a demobilization plan with specific instructions as to when, where, and how personnel are relieved is a task suited for the Planning Section.

SAFETY

Another concern is safety. Safety is the first operational priority. On a defensive operation, there is more time available to develop a comprehensive safety plan. Multiple safety officers may need to be assigned in order to address all the hazards present. The safety team would make an assessment of the hazards and risks present on the fireground. Any immediate life safety hazards would be addressed during the walk-around. After the assessment is made, the safety team would discuss their findings and make recommendations to eliminate or reduce the exposure to the hazards. After the plan is approved by Planning and the incident commander, the Operations Section is briefed, and the plan is implemented and enforced.

EXPOSURES

Exposures are a fourth concern. They constitute a major complexity of defensive fire attack. Protecting the adjacent and downwind structures can be a major undertaking. Because fire brands have been known to travel long distances, the Exposure Sector/Division may be operating remote from the main fire. They may be assigned to external exposures, such as the building next door or downwind, a block away. They may also be assigned to internal exposures, such as an uninvolved floor above or below the fire.

The Exposure Division supervisor has numerous responsibilities. Some of these include the safety and accountability of all assigned personnel and requesting additional resources to handle all tasks necessary to complete the objectives. The Exposure Division supervisor is also responsible for updating the Operations and Command personnel on the status of the exposure buildings and the progress of the protection efforts.

In a defensive fire attack, officers must have a realistic anticipation of the incident magnitude and potential. To assist in the evaluation, ask yourself these questions:

ANTICIPATING INCIDENT MAGNITUDE AND POTENTIAL

- How big can this thing go?
- Are we having any effect on the fire?
- How long will it take us to retreat from this position if we are forced to?
- Where is the safest place to retreat?

Summary

The most important skill you having in keeping your crew safe and alive is size-up. Knowing when to go marginal and when to go defensive is critical. The second most important skill is understanding and identifying fire behavior. Know your enemy!

This is war! It is a battle. In every battle there must be casualties. The only acceptable outcome is for the enemy to sustain the fatal strike. Killing or injuring your personnel is an unacceptable outcome. In order to defeat the fire, you must understand the enemy's moves, strikes, and power. You must also know your own capabilities and limitations. Safeguard your personnel, and provide for firefighter safety and personnel accountability first. Always respect the collapse zone, and remember that every evacuation order is to be obeyed, every time.

Note: The Command Worksheet on the next page is for use with the student exercises that follow. This sheet may be duplicated as needed.

INCIDENT INFO

Incident Name ______________________

Time/Alarm	Address	Unit No.	Inc. #	Date
Time/Scene	Occupancy – Size/Type/Acres		Property Loss	
1st. Scene	Occupant	Telephone	Contents Loss	
Control	Owner	Telephone	Civ. Inj.	F.D. Inj.

RESOURCE STATUS

Company	A/S	Assignment

COMMAND WORKSHEET

☐ INITIAL INCIDENT ACTION PLAN

1 ☐ Offensive ☐ Marginal ☐ Defensive

2

LIFE SAFETY OBJECTIVES	INCIDENT CONTROL OBJECTIVES	PROPERTY OBJECTIVES
☐ ☐	☐ ☐	☐ ☐
☐ ☐	☐ ☐	☐ ☐
☐ ☐	☐ ☐	☐ ☐

3 # Co's Req'd ____ # Co's Req'd ____ # Co's Req'd ____

Plus R.I.C. ____ → TOTAL # CO'S ____ # ALARMS ____

4 Communications Plan ☐ Command ☐ Tactical ☐ Air to Ground Tac

5 Level II Staging Area Location

Rescue		Vent	
Yes	Clear	Smoke	Clear

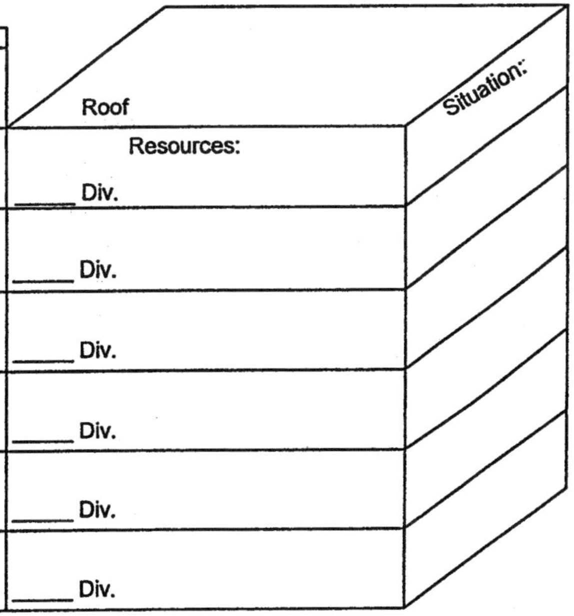

TACTICAL BENCHMARKS

- ☐ Fire attack
- ☐ Water supply
- ☐ Primary search
- ☐ All clear
- ☐ Utilities
- ☐ Ventilation
- ☐ Back-up lines
- ☐ Ext check / exposures
- ☐ Salvage
- ☐ Secondary search
- ☐ All clear
- ☐ Progress reports
- ☐ Rescue laddering
- ☐ Sprinkler support
- ☐ Breathing / lighting support
- ☐ Rehab
- ☐ Traffic control
- ☐ P.I.O. / media notification
- ☐ Under control
- ☐ Investigator
- ☐ Utility co's
- ☐ Building rep
- ☐ Building and safety
- ☐ Release plan

Student Exercise 6-1

In this exercise, the student is playing the role of the company officer on the first-arriving apparatus. The student should complete the following:

1. Receive the initial dispatch information.
2. View the written and graphical prefire plan information.
3. Conduct a size-up from the initial view upon arrival.
4. Transmit a radio size-up.
5. Develop an ICS chart for the initial incident organization.
6. Assign all first-alarm companies by radio.

Dispatch Information

You have just been dispatched to a commercial structure fire. From the address, you realize that this suburban fast food restaurant was built within the past five years. The building is a one-story, commercial restaurant and drive-through. It is constructed of wood frame, lightweight trusses, and an asphalt /composition roof with a mansard on three sides. The time of alarm is 2306 hours. The temperature is 35°F, and the humidity is 35%. Winds are from the west at 11 mph.

"Battalion 4 … commercial structure fire. Attention, Battalion 4 units and stations. A commercial structure fire, fire reported in the restaurant, 77 Cypress Street. 77 Cypress. Cross street: Seventh Avenue. Respond on channel 3 at 2306 hours."

"Battalion 4 units are responding."

"Emergency World received. First unit is on-scene."

Arrival Information

Upon arrival, you find fire and smoke showing from the interior of the restaurant. The entire dining area is involved, and the fire has just vented out through the front door. The fire has not vented the roof.

You are responding with your department's first-alarm assignment. The hydrants are spaced according to your department's commercial subdivision ordinance. The instructor will advise on hydrant location.

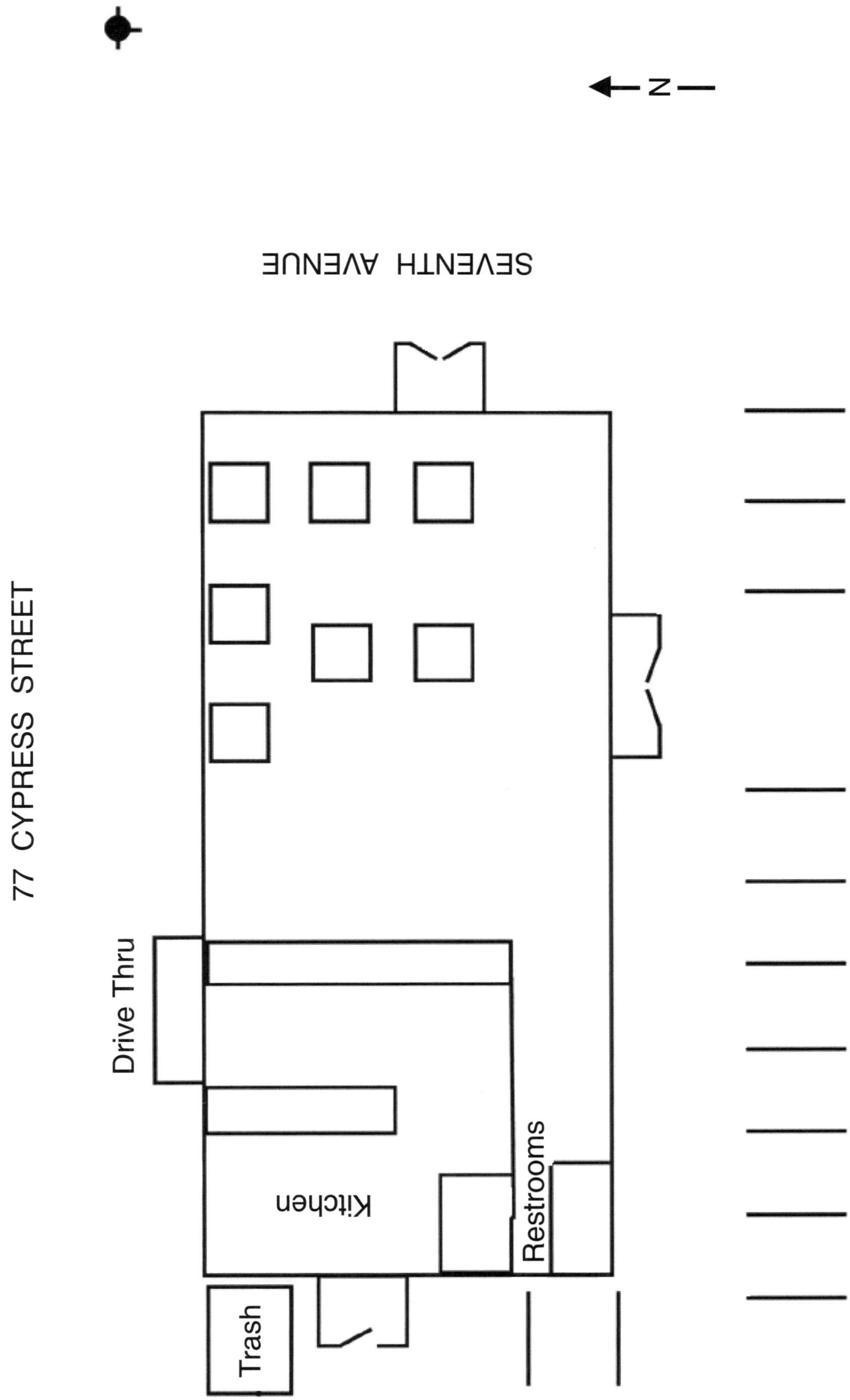
N
SEVENTH AVENUE
77 CYPRESS STREET
Drive Thru
Kitchen
Restrooms
Trash

Student Exercise 6-2

In this exercise, the student is playing the role of the company officer on the first-arriving apparatus. The student should complete the following:

1. Receive the initial dispatch information.
2. View the written and graphical prefire plan information.
3. Conduct a size-up from the initial view upon arrival.
4. Transmit a radio size-up.
5. Develop an ICS chart for the initial incident organization.
6. Assign all first-alarm companies by radio.

Dispatch Information

You have just been dispatched to a commercial structure fire. From the address, you realize that this downtown commercial strip or mercantile building was built in the 1920s. The occupancy is a taxpayer with a small apartment above the retail space below. The bottom occupancy is a tax consulting office. The building is a two-story, commercial strip center. Other occupancies are a print shop, a pawn shop, a drugstore, a furniture store, and an electronics shop. The building is constructed of unreinforced masonry with a conventionally constructed wooden, flat roof. The time of alarm is 0827 hours. The temperature is 55°F, and the humidity is 26%. Winds are from the west at 8 mph.

"Battalion 3 ... commercial structure fire. Attention, Battalion 3 units and stations. A commercial structure fire, fire reported in the tax office, 152 First Street. 152 First Street. Cross street: West Main Street. Respond on channel 2 at 0828 hours."

"Battalion 3 units are responding."

Arrival Information

Upon arrival, you find fire and smoke showing from the second-floor apartment. The smoke is visible from one window on the front and one from the side. The fire has not vented the roof.

You are responding with your department's first-alarm assignment. The hydrants are spaced according to your department's commercial subdivision ordinance. The instructor will advise on hydrant location.

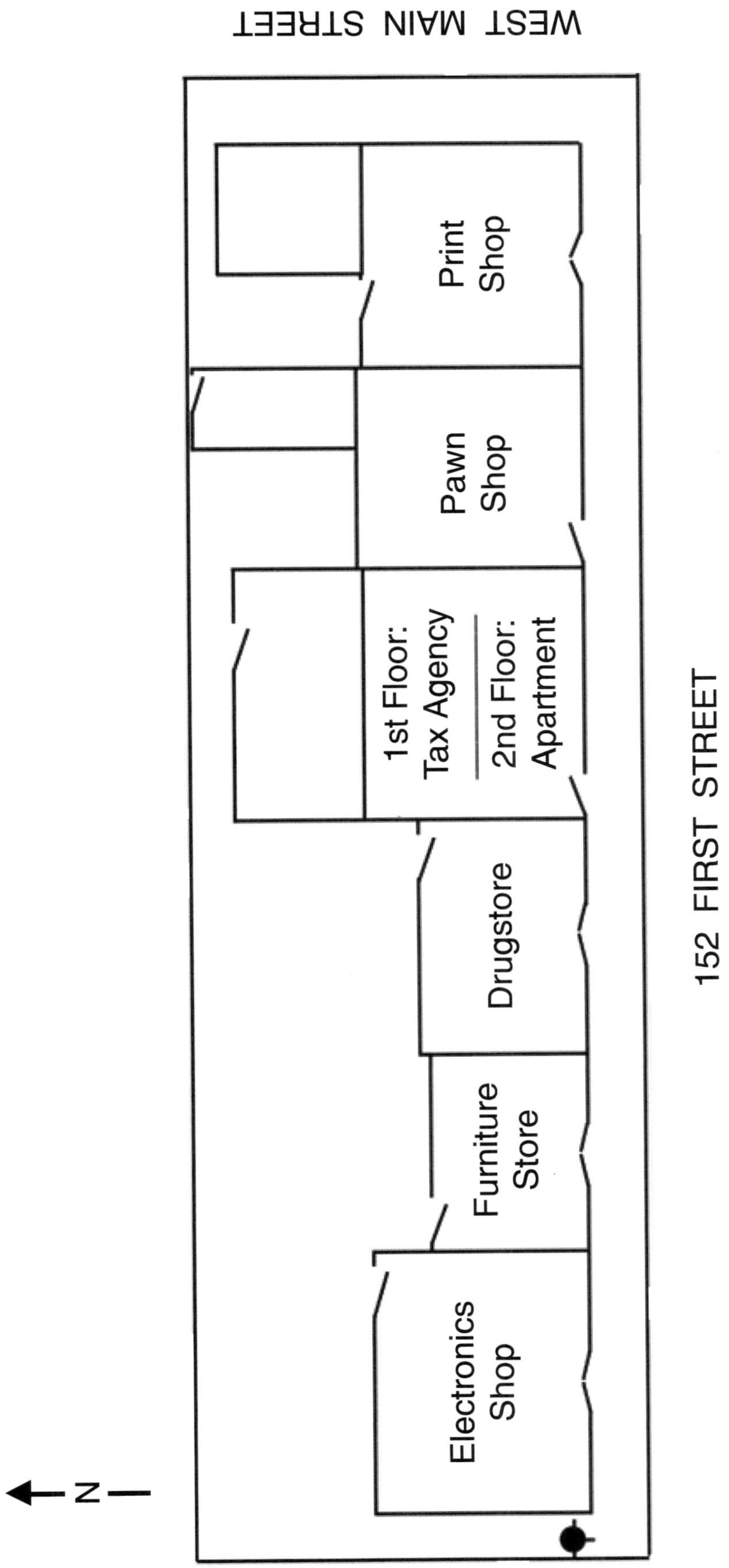
WEST MAIN STREET
Print Shop
Pawn Shop
1st Floor: Tax Agency
2nd Floor: Apartment
Drugstore
Furniture Store
Electronics Shop
152 FIRST STREET
N

Student Exercise 6-3

In this exercise, the student is playing the role of the company officer on the first-arriving apparatus. The student should complete the following:

1. Receive the initial dispatch information.
2. View the written and graphical prefire plan information.
3. Conduct a size-up from the initial view upon arrival.
4. Transmit a radio size-up.
5. Develop an ICS chart for the initial incident organization.
6. Assign all first-alarm companies by radio.

Dispatch Information

You have just been dispatched to a commercial structure fire. From the address, you realize that this older downtown building was built in the 1930s. The occupancy is a commercial florist and houses several other small office businesses. The building is a one-story, multi-tenant, commercial building. It is constructed of unreinforced masonry block walls and a bowstring truss roof. The roof covering is asphalt/composition. The time of alarm is 0931 hours. The temperature is 65°F, and the humidity is 21%. Winds are from the northeast at 2 mph.

"Battalion 2 ... commercial fire alarm. Attention, Battalion 2 units and stations. A commercial fire alarm, 200 East Main Street. 200 East Main Street. Cross street: Broadway. Respond on channel 2 at 0931 hours."

"Battalion 2 units are responding. The alarm company advises they received a general alarm followed by a water flow."

"Emergency World received. First unit is on-scene."

Arrival Information

Upon arrival, you find heavy fire and heavy black smoke showing from two sides of the structure. Fire and smoke are pushing out of all available roof vents under pressure. There is fire visible from four windows on the west side. The fire has not vented the roof.

You are responding with your department's first-alarm assignment. The hydrants are spaced according to your department's commercial subdivision ordinance. The instructor will advise on hydrant location.

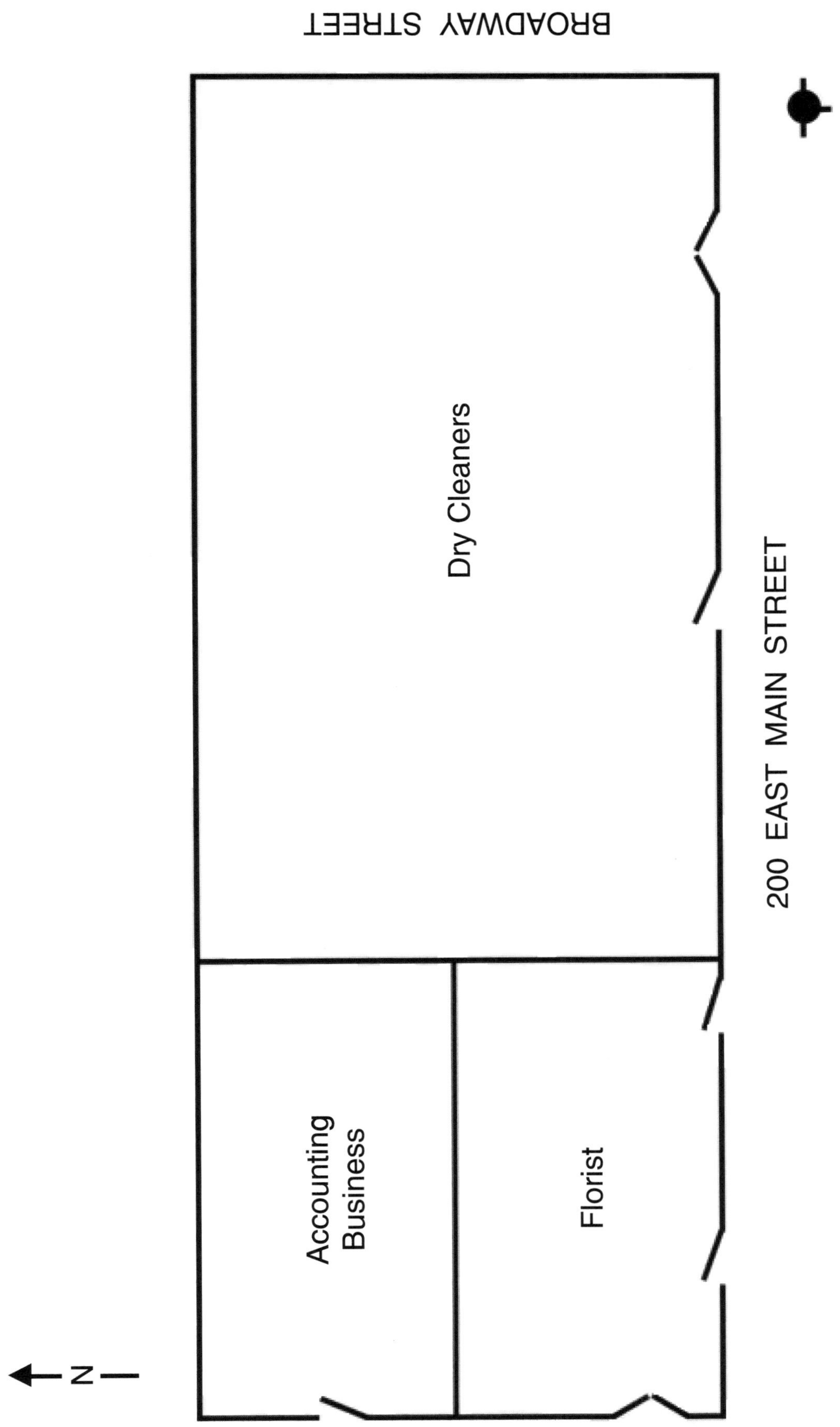
BROADWAY STREET
Dry Cleaners
200 EAST MAIN STREET
Accounting
Business
Florist
N

Index

A

accountability system, 32, 50, 53-55, 93, 141, 142, 146
acetylene, 2
action plan, 24, 26, 27-29, 37, 40, 42, 43
aerial devices, 72, 79-80, 84, 144
air monitoring, 59, 113, 115, 118, 123
"all clear" benchmark, 51, 74, 85, 140
American Conference of Governmental and Industrial Hygienists, 10
apparatus placement, 29-30, 37, 67, 72, 84
appliances, 143
arcing, 5
assessment, 8, 14, 58, 59, 61, 68, 78, 91, 145
    See also size-up
assuming command, 26
atmospheric monitoring, 59, 113, 115, 123
attack lines, 67-68, 80, 104, 143
attack monitors, 139, 141, 144
attic ventilation, 111

B

backdraft, 13-14, 75, 93, 95, 105
backup lines, 18, 19, 90
backup team, 88, 90
below-grade conditions, 118
benchmarks/benchmarking, 41, 42, 51, 66, 74
BLEVE, 6
boiling point, 3
"box" dynamics, 70-71
British thermal unit (BTU), 2
building construction types, 14-16, 19, 78, 82

C

calorie, 2
carbon dioxide, 10, 11, 17, 94
carbon monoxide, 8-9, 11, 12, 13, 59, 78, 94, 106, 107, 113, 115, 123
carboxyhemoglobin, 9
carcinogens, 8, 95
casement windows, 108
Celsius, 2
chemical chain reaction, 6, 12, 17, 18
chemical heat sources, 5
churning, 99
Class A fires, 7, 17
Class B fires, 7, 17
Class C fires, 7, 17
Class D fires, 7, 17
closet/vault ventilation, 110
collapse zone, 30, 50, 60, 139, 142, 144, 146
combustible gas meters, 3
command modes, 25-27
command post, 26
command presence, 21-23, 33
command worksheet, 39, 40, 41, 42-43, 47
communications, 32, 43-45, 50-52, 92-93, 117, 142
company integrity, 54, 67, 82
compression, 5
conduction, 8
confined space ventilation, 112-115
confining the fire, 28, 30, 64, 65-66
construction time periods, 16
construction types, 14-16
convection, 8, 12
coordination, 26, 31, 45, 82, 83, 86, 87, 95, 106, 107, 144-145
cord reels, 122
coverage, 145
crew rotations, 55
    See also relief

D

decay stage, 12, 13, 94
decomposition, 6
defensive strategy, 27, 31, 137-146
defensive strip, 76-77, 81
deodorants, 125-126
deployment examples, 36
dielectric heating, 5
direct attack, 64, 105
disposition options, 26
double-hung windows, 107
dual-pressure automatic nozzles, 69-70, 143
ducts for ventilation, 107, 113, 118, 119

E

egress, 24, 36, 73, 75, 79, 84, 90, 91, 96, 102, 104, 106, 107, 113, 116, 142
electric fans, 102, 105, 106, 107, 113, 123
electrical fires, 7, 17
electrical heat sources, 5
elevated fire streams, 72, 80, 84, 139, 144
embers/fire brands, 80, 145
engine company operations, 63-70
equipment use instructions, 121-122
explosion-proof motors, 98, 102, 107, 113, 119, 123
explosive limit, 2
Exposure Division, 145
exposures, protecting, 30, 31, 80, 84, 140, 143, 145
extension check, 28, 72, 78, 80, 81
extension cords, 122
exterior protected combustible construction, 14, 15
extinguishing a fire, 17-18, 28, 63, 64, 65, 66, 79, 80, 82, 93, 94, 143

F

face-to-face communications, 32, 43, 44, 45, 142
Fahrenheit, 2
fans, 77, 78, 94, 96, 98, 99-111, 113, 115, 117, 121, 123, 124-125
fatalities among firefighters, 49-54, 67, 70, 82, 137, 138, 142
fire behavior terms, 2-3
fire, by-products, 8
See also products of combustion
fire, definition, 3
fire extension, 70, 71, 76, 77, 84, 91
fire, extinguishing, 17-18, 28, 63, 64, 65, 66, 79, 80, 82, 93, 94, 143
fire flow formulas, 18-19, 140
fire gases, 19, 28, 59, 73, 74, 76, 93, 94, 96, 102, 103, 105, 116
fire investigation, 66
fire movement, 7-8
fire prevention, 7, 8, 138
fire resistance, 14
fire resistive construction, 14, 15
fire stages, 11-12
fire tetrahedron, 4-6, 12, 17, 59
fire types, 7
"firefighter creep," 142
fireground accountability, 67, 91, 93
See also accountability systems
Fireground Command System, 52, 53
first-in company officer responsibilities, 23-31
flammable gases, 2, 8
flammable liquids, 2, 7, 8, 17, 139
flammable metals, 7, 17
flammable range, 2
flash point, 3
flashover, 13, 61, 66, 67, 75, 82, 84, 93, 95, 139
focus, 51, 52, 53, 57, 88, 89, 90, 92, 93, 140
fog streams, 94, 96, 119-121
forcible entry, 57, 71, 72-73, 80, 81
free-burning stage, 12, 13, 67, 94
free-lancing, 31, 52, 53, 55, 142
friction, 5
fuel, 6, 12, 17, 18, 65, 93, 120

G

gasoline powered fans, 78, 96, 98, 99, 102, 105, 113, 115, 117, 123, 124
generators, 121-122

H

handlights, 57, 58
handlines, 68, 143
hazards, 60
heat, 2, 4-6, 8, 12, 13, 14, 19, 73, 76, 77, 78, 79, 84, 93
heat hole, 76, 78, 81
heat/temperature reduction, 17, 95, 119, 120
heat transfer, 7-8, 19, 76
heavier-than-air gases, 107, 113, 114, 115, 118, 123
heavy timber construction, 15, 16
high-rise buildings, 116-117
hoselines, 67-68, 90, 98, 143
hydrogen chloride, 9-10
hydrogen cyanide, 10
hydrogen gas, 8

I

IDLH conditions, 89, 90
ignition temperature, 3, 4, 5, 6, 12, 17, 95, 120
incident action plan, 24, 26, 27-29, 37, 40, 42, 43
incident command organization, 32, 34, 35, 46, 85, 144
Incident Command System, 31-41, 52-53
incident communications systems
See communications
incident/fire control issues, 24, 28, 29, 31, 42
Incident Management System, 52, 53
incipient stage, 11-12, 94
indirect attack, 65
ingress, 104, 106, 113, 116
initial incident action plan, 24, 26, 27-29, 37, 40, 42, 43
injuries among firefighters, 49-54, 144
inside operations, 80, 81
inside/out attack, 66, 84
internal room ventilation, 111, 119
investigation command mode, 26
Iowa State University Formula, 18-19

L

ladders, 71, 75, 79, 80, 81
large-diameter hoselines, 64, 67, 139, 143
leakage current heating, 5
Level 1 staging, 37, 38
Level 2 staging, 37, 38, 43
life safety issues, 24, 28, 29, 30, 42, 145
light, as by-product of fire, 8
lighter-than-air gases, 113, 114, 118
lighting, 57, 59, 74
lightweight era, 16
locating the fire, 64, 65, 105
"loss stopped" benchmark, 51, 66
lost personnel, 50, 55, 57, 59, 60, 66-67, 82, 141

M

manholes, 113, 118
manpower, 93, 96, 116, 117
See also staffing
marginal strategy, 31, 140-141
master streams, 68, 84, 139, 144
mayday, 59, 142
mechanical heat sources, 5

mechanical ventilation, 59, 94, 96-101, 120
microphones, 44, 51
monitoring the atmosphere, 59, 113, 115, 118, 123

N

National Fire Academy, 19, 52, 62
National Fire Academy Formula, 19, 140
natural ventilation, 94, 96
negative pressure ventilation, 94, 106-111
neutralizers, 125
NFPA 1500, 57
NFPA 1964, 69
NFPA 220, 14, 15
nitric oxide, 10
nitrogen dioxide, 10-11
"no all clear," 85, 140
non-combustible construction, 14, 15
nozzles/nozzle pressure, 68-70, 143
NPV
See negative pressure ventilation
nuclear heat sources, 6

O

Occupational Safety and Health Administration, 56
offensive strategy, 27, 30, 83-126
operational retreat, 141-142
Operations Section, 33, 145
oriented search, 85, 86-87, 91, 92-93
OSHA, 56
outside operations, 80, 81
outside/in attack, 66, 84
overhaul, 65, 66, 78, 79, 94, 96, 98, 102, 107, 120, 126
"ownership" of address, 29-30, 72, 84
oxidation, 3
oxidizers, 4
oxygen, 4, 9, 11, 12, 17, 59, 94

P

PAR
See personnel accountability report
passing command, 27, 45
personnel accountability report, 55
personnel accountability system, 50, 53-55, 62
phosgene, 11
physical fitness, 22-23, 50, 61
placement of apparatus, 29-30, 37, 67, 72, 84
Planning Section, 55, 61, 144, 145
portable generators, 121-122
positive pressure ventilation, 77-78, 94, 96, 102-106
post-World War II construction, 16
PPN method, 51
PPV
See positive pressure ventilation
pre-1933 construction, 15, 139
"preconnect syndrome," 68
predesignated seat assignments, 80-82
prefire planning, 22, 98, 116, 117, 138, 139
pressure, 64, 69
pre-World War II construction, 16
primary search, 73, 84, 85, 88-89, 140
products of combustion, 13, 76, 77, 78, 91, 93, 94, 95, 105, 115
progress reports, 40-41
property damage, 31, 66, 78, 79, 93, 96, 102, 107, 126
property salvage issues, 24, 28, 29, 31, 42
See also salvage
protecting exposures, 30, 31, 80, 84, 140, 143, 145

R

radiation, 8, 12
radio channel designations, 43-45
radio communications, 32, 43-44, 50, 51, 54, 57, 59
radio size-up, 24-25, 27, 44
rally points, 142
rapid intervention teams, 29, 43, 57-59, 62, 88
recirculation, 109
reducing temperature, 17, 95, 119, 120
relief, 145
See also crew rotations
relief valves, 6
rescue, 30-31, 83, 89-90, 93, 94, 95, 105, 117, 140, 141
Rescue Group, 89
resistance heating, 5
resource status tracking, 26, 39-40, 42, 46, 145
resources, 19, 24, 25, 26, 27, 28, 29, 31, 32, 36, 37, 45, 51, 87, 88, 90, 140, 141
RESTAT
See resource status tracking
risk analysis/assessment, 50, 60-62, 91
risk management, 60-62
risk matrix, 61
RIT
See rapid intervention teams
rope, 57, 58, 86

S

saddle stack fan placement, 104
safety, 32, 66-67, 70, 82, 86, 87, 88, 91, 92, 102, 107, 113, 123, 145
salvage, 72, 78-79, 81, 83, 126
See also property salvage issues
SCBA, 58, 59, 61, 89
search and rescue, 71, 73-74, 80, 81, 89
search operations, 28, 65, 83, 85-93, 140, 141
seat assignments, 80-82
secondary search, 74, 84, 85, 140
securing the scene, 65, 66
selective ventilation, 111
series method of fan placement, 100
SITSTAT
See situation status tracking
situation status tracking, 26, 39, 42, 46, 144
size-up, 24-25, 26, 27, 44, 58, 68, 87, 139, 141, 146
smoke, 8, 11, 12, 13, 14, 19, 70, 73, 74, 76, 77, 78, 84, 93, 94, 95, 96, 98, 102, 103, 116, 117
smoke removal, 29, 102, 117, 125, 126
SOPs
See standard operating procedures
span of control, 32, 35, 37, 53, 142
SPIN format, 41
spontaneous heating, 5
staffing, 27, 50, 56, 62, 79, 80, 81, 87, 89, 138, 139
See also manpower
staging, 37-38, 58
standard operating procedures, 37, 38, 52, 98, 145
standard search, 85, 86, 87
standpipe systems, 69, 72, 80
static electricity, 5
static water loads, 28, 59, 71, 75, 78, 79
strategies, 21, 27, 30
See also defensive strategy; marginal strategy; offensive strategy
stress, 22, 23, 59
structural collapse, 14, 27, 60, 66, 67, 76, 78, 82
structural integrity, 70, 76, 78
sudden hazardous events, 66-67
sulfur dioxide, 94

T

tactical benchmarking, 41, 42, 51
tactical deployment, 36
tactical objectives, 21, 27, 28, 29, 30, 31, 32, 33, 39, 42, 45, 47
team search, 85, 86
temperature emergencies, 119
temperature/heat reduction, 17, 95, 119, 120
thermal imaging cameras, 58, 65, 91
TICs
  See thermal imaging cameras
"toolbox" approach, 33, 34, 35
tools, 57, 58, 91, 143-144
toxic gases, 10, 11, 12, 59, 73, 77, 93
transfer of command, 43, 45-47
trapped personnel, 29, 50, 55, 57, 59, 60, 66-67, 82, 91, 138, 141
truck company operations, 70-80
trussed construction, 15, 16, 139
two-in/two-out rule, 56, 86
Type I construction, 15
Type II construction, 15
Type III construction, 15
Type IV construction, 15
Type V construction, 16

U

"under control" benchmark, 41, 46, 51, 55, 66, 74
unreinforced masonry construction, 15, 16, 60, 139
utility control, 71, 74-75, 80, 81

V

vapor density, 113
vapor pressure, 3
V-attack procedure, 100, 104
vault/closet ventilation, 110
ventilation, 13, 14, 28, 59, 71, 74, 75-78, 80, 81, 82, 83, 85, 93-126
ventilation decision tree, 97, 112
venturi effect, 103, 109
vertical ventilation, 14, 76, 77, 78, 81
visibility, 11, 73, 74, 76, 93, 94, 102, 107

W

water pressure, 64
water reactive materials, 5, 7
water supply, 64, 143
water triangle, 63-64
water vapor, 11, 94
wind direction/speed in ventilation, 98, 100-101
wood frame construction, 15, 16